Daniel J. D. Sullivan, Eric J. Carleton

Failure Analysis

I0051003

Also of Interest

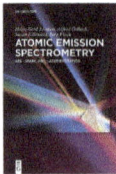

Atomic Emission Spectrometry.
AES – Spark, Arc, Laser Excitation
Heinz-Gerd Joosten, Alfred Golloch, Jörg Flock and
Susan Killewald, 2021
ISBN 978-3-11-052768-1, e-ISBN 978-3-11-052969-2

Chemical Analysis in Cultural Heritage
Luigia Sabbatini and Inez Dorothé van der Werf (Eds.), 2020
ISBN 978-3-11-045641-7, e-ISBN 978-3-11-045753-7

Electrophoresis.
Theory and Practice
Budin Michov, 2020
ISBN 978-3-11-033071-7, e-ISBN 978-3-11-033075-5

Rubber Analysis.
Characterisation, Failure Diagnosis and Reverse Engineering
Martin J. Forrest, 2019
ISBN 978-3-11-064027-4, e-ISBN 978-3-11-064028-1

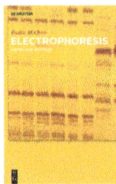

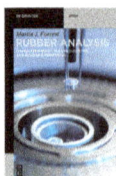

Daniel J. D. Sullivan, Eric J. Carleton

Failure Analysis

High Technology Devices

DE GRUYTER

Authors
Dr. Daniel J. D. Sullivan
EAG Laboratories
1708 McCarthy Boulevard
Milpitas 95035
United States of America
djdsullivan@gmail.com

Dr. Eric J. Carleton
EAG Laboratories
810 Kifer Road
Sunnyvale, CA 94086
United States of America
eric.carleton@gmail.com

ISBN 978-1-5015-2478-3
e-ISBN (PDF) 978-1-5015-2479-0
e-ISBN (EPUB) 978-1-5015-1647-4

Library of Congress Control Number: 2022941691

Bibliographic information published by the Deutsche Nationalbibliothek
The Deutsche Nationalbibliothek lists this publication in the Deutsche Nationalbibliografie;
detailed bibliographic data are available on the internet at http://dnb.dnb.de.

© 2022 Walter de Gruyter Inc., Boston/Berlin
Cover image: matejmo/iStock/Getty Images Plus
Typesetting: Integra Software Services Pvt. Ltd.
Printing and binding: CPI books GmbH, Leck

www.degruyter.com

This book is for all those that toil in the labs, the fabs, and the field figuring out what went wrong so that technology may progress.

Dan Sullivan – Dedicated to my wife Christie and the doctors and staff at UCSF without whom I would not be here to write this book.

Eric Carleton – Dedicated to my parents, Boots and Barbara Shumway, and the medical staff at NBBH, Billings Clinic, and the Mayo Clinic without whom I would not be in a position to write this book.

It is the intentions of the authors to donate any royalties they receive for the sale of this book. The donated proceeds will go to the Multiple Myeloma Foundation (https://www.myeloma.org/) and the Epilepsy Foundation (https://www.epilepsy.com/) to help increase outside awareness, to fund research, and pursue cures.

We are keeping the movie rights money. We have always thought that a sitcom called "Life in the Lab" based on craziness of high-tech labs could be good and accidently educational.

Contents

Chapter 1
Introduction to FA

The world of failure analysis is large. We hope this work will encourage new comers, give a broader range of knowledge to the journey men/women, and give some familiar and funny examples to the old timers. In this book, we will attempt to cover only the corner of the FA world considering high tech devices and materials. So, we will leave out buildings, bridges, dams, and other large objects, as well as the mysterious area of software. The failure investigation investigating why a system fails (a phone, television, smart watch, or any other device made up of several components) is different than the FA of a single part (resistor, chip, tube), although these are in turn made up of smaller parts themselves. No matter the subject of an investigation, the goal of FA remains unchanged. To find out what caused the failure: the root cause. Even finding the root cause is rarely the end purpose. Knowing the cause allows the factory to change the processes used to make the part, to repair or clean the tools that contaminated the process, or the engineers can perform a design change to improve the end product and greatly reduce or eliminate the failure. Occasionally, the root cause determination is for the purpose of assigning responsibility in litigation/ criminal actions or other situations when the improvement of the device is not the driving purpose. This is why failing parts are sometimes referred as "golden units", in that they are rare and valuable. They offer a chance to better understand and improve a device or process or decide who gets the gold in litigation matters. These parts are also sometimes referred to as "incident units"; however, this nomenclature is often discouraged by attorneys as it sounds bad.

Figure 1.1: Image of an ant and a VCSEL (vertical cavity surface emitting laser). The path to the discovery of a failure can sometimes lead you to very unexpected events, like the classic "bug" in the system. Image by Jason Tan.

https://doi.org/10.1515/9781501524790-001

Failure analysis often accompanies reliability testing of products. The reliability tests are done to determine if a part will last as long as it is intended or determine the failure rates of parts when they are in the field. The warranty people are very interested in these answers. In some instances, the reliability or stress tests are done until the parts fail, this allows inspection of the types of failures that occur thus identifying the weaknesses of the material or design.

In this book, the parts are typically three types: wafers, devices, and PCB systems. Of course, some of the processes here can be used on any physical object from steel tubes to tennis shoes.

Figure 1.2: Image of wafer, packaged device, and PCB system. Image by Dan Sullivan.

It is also common that customers require a qualification of new parts or whenever a change is made to a device that alters its fit, form, or function. This often requires a written notification to the customer along with a requalification of the device. I have heard many arguments over what constitutes a real change and if a part can be requalified by extension in that it is substantially like another part that has already been tested and passed. Most of these arguments are near-sighted attempts to speed up the shipping of parts to hit a monthly goal and regularly come back to haunt the company later. In larger companies, factory processes such as "Copy Exactly!" are utilized to control and mitigate these issues.

In a Material Review Board (MRB), I have heard a director of packaging say that only the size, the materials for underfill, and the layout were changed, so the device was essentially the same. This was an attempt to avoid having to requalify the device. In this case, I am proud to say he was overruled.

In these qualification tests, there are usually several standard tests that just change the temperature, humidity and the bias the parts experience. Elevated temperature (HTOL) and humidity (DH) as well as swings in temperature (TC) are commonly used to test and qualify ICs. The duration and severity of the tests depends on the industry involved and the desired rating of the device as well as the customer demands. These tests can take months to complete and involve hundreds of parts, so this is not cheap. When a part fails in a qualification, the FA must be very thorough and fast. There can be millions of dollars at stake so no pressure.

The specific test in which the failure occurs in must be known to the FA team. The main question will be: "Is this a legitimate failure or not?" Knowing the test the part experienced will help direct the FA and rule out some failure modes as not possible. If it is a legitimate failure then the qualification may fail requiring a repeat of the qualification test that failed or even redesign of the product. These repeat tests cost a lot of money and even worse will delay the product time to market or possibly result in the cancelation of the project.

If the failure is not legitimate then that sample may be put aside and one of the extra parts (there are always a couple of extra parts run) is used instead. There will be a lot of pressure brought to bear to have this false failure as the conclusion of all FAs in a qualification. You must be able to resist such pressure and be impartial as an FA person. If you ever compromise your judgment and provide false results or conclusions you are done. No one can believe you going forward. Even if you save the project or company for a while by providing false evidence you will be damaged and unlikely to have any sort of future in FA. Remember you are not the cause of the failure, you are there to provide the best analysis and truthful results so others can make informed business and safety decisions. In cases that involve litigation, or the potential thereof, properly reporting so-called negative results can assist in avoiding additional harm and damages as well as save you from exposing yourself personally in any legal matters.

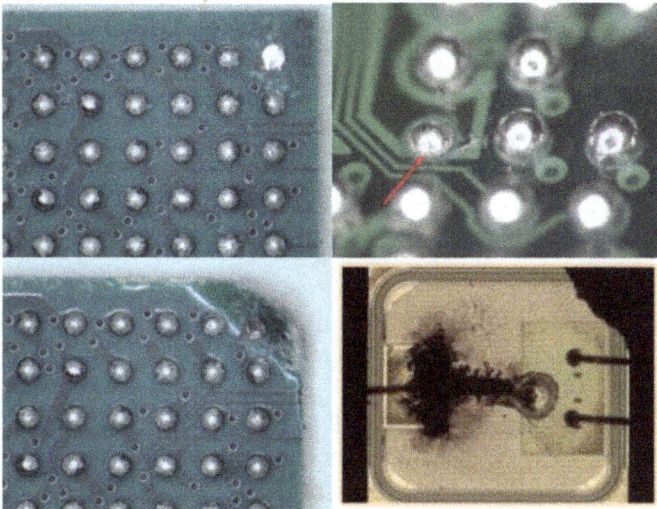

Figure 1.3: Examples of false failures found during qualification testing. Sheared solder ball, insufficient solder, package damage, and ESD damage. Image by Dan Sullivan.

The other major source for FA parts is field returns. They are much less desirable as this means a customer has experienced a failure of your device. While less desirable, they are also very valuable to the FA lab and the company as a whole as they offer an insight into failures under real operation conditions and even though the customer may be unhappy that they have a failure, the FA process offers a chance to show them that you take the issue seriously, that you are professional and competent as a company. Hopefully, once the FA is complete, you will share the information with the customer and it will lead to improvements in design, manufacture, or operations instructions. This should improve the product and the relationship with the customer.

Remember that no one wants to pay for FA or reliability testing and many companies try to get out of it or do the least they can. They are idiots. When done properly, FA and reliability help improve yields, reduce warranty returns, and provide a good company/product reputation. Everyone understands that things can go wrong. Addressing issues with interest and openness will help the situation when they occur.

FA is a tool for companies to use to improve their products, cut costs, and ensure customer satisfaction and safety. When properly staffed and funded, FA is a benefit to the company for controlling costs and enhancing reputation. Failure analysis, and for that matter all testing, can be done in house or through external vendors. The place the FA is done does not matter as long as it is done properly and with integrity. Any company that wants to make decisions based on false evidence or blindly by ignoring available opportunities to understand the issues is not long for this world.

I have been asked many times if it is better to have FA in house or just use outside labs. I have run both types of labs and the answer is of course – it depends. If you have a cutting edge or defense-related technology then security is a large concern and in-house labs have some advantages here as the devices never have to leave the custody of the company. Before entrusting samples to an outside vendor an audit or at least a visit to the site is advisable. Do they have a secure location for sample storage when not being worked on? Do they have a secure server system for the results and reports? What is their policy on sharing results? What is the process for screening the individuals that have access to your parts? All these questions should be answered to your satisfaction before you move on to the next questions. Additionally, a non-disclosure agreement (NDA) is almost always a requirement before any work or sharing of IP details is begun.

One additional detail on military devices is ITAR (International Traffic in Arms Regulations). There are rules as to who may work on or even be in the room when such devices are present. If you will be working on ITAR parts for FA, you and your company need to understand the rules and processes that must be followed. I do not want any of the readers of this book to end up in federal prison. There are many specifications for testing and analysis, but this is the only one I know of that can carry felony charges. You will have to learn what a "US person" means.

Turnaround times (TAT) and cost are the next big issues. I have yet to meet the person who does not need their results very quickly – yesterday or the day before is fine. If you have ways to triage work coming into the lab that is best. If every curious Karen's work must be done and there is no ranking of importance you wind up with FIFO, first in first out, which is a terrible system, only slightly better than the squeaky wheel system. The best triage systems are based on cost: no one pays for unimportant work. For internal labs where cost is not possible to use, a limited number of slots allowed in the lab for each "customer" division works well. It is even better if they have rankings for their slots. So, if VP Joe wants an FA done immediately, they use their #1 slot and if that is currently occupied, they replace the other request with this more important one. At one lab I ran, we provided laminated cards with 1, 2, 3, and so on printed in large font on both sides which were handed to the VP of each department that provided funding to the lab. Each had the name of the department and a color making it easy to identify. The total number of cards was based on each person in the lab working on five jobs at any given time. This number was based on how many fingers I have on one hand and is somewhat arbitrary, but something between 3 and 10 projects is the most a person can juggle reasonably. The number of cards/projects each division received was based on the percentage of the FA budget that division provided. This worked remarkably well. Unfortunately, I have also run labs in companies that believed every job submitted needs to be done and nothing has priority. If you guessed, I had constant calls about the newest hot job that had to go first, you are psychic. With no allowed method of triage, it became a game of which VP called me last. This was a waste of my time and theirs.

Using something like the ranking cards, the departments are able to decide which of their work requests should go first and what is not important enough to get a slot. This also makes very clear to all parties the resources available. The lab resources for each department is directly related to the support the lab receives. This made discussion of the budget for the FA lab much clearer for all involved. If they added support for an additional headcount, then five more cards would become available and the departments that put in the support would get the cards, easy.

The costs are always an issue. The cost of doing work internally is of course the cost to set up the lab: space, people equipment, upkeep, and usually some work with the fire department and health department. This is a large obstacle when first setting up a lab as opposed to smaller expansions to an existing lab. When deciding to set up a lab or not or adding substantial staffing or capital equipment an ROI (return on investment) sheet should be done.

The ROI is fairly easy to do. Determine the full costs of what you are adding, say a new X-ray imaging machine that will take up a 10 ft by 10 ft area in the lab and require a technician to run. Get the tool costs from the tool vendor, say $100,000 in this case. Say the monthly cost of the lab space all in, get the actual number from your facility manager, is $2 per square foot. Now the technician you are hiring you can estimate from current staff salaries or HR can help you out here, let's say it is $50,000 per year. If the upkeep is minimal, no major process gases or other materials, then along with power is only $200 per month. So, amortizing the tool cost over 5 years, good luck with that as it is quite often 3 years or even 18 months in some companies, a total yearly cost is $20,000 for the tool plus $2,400 for the space plus $50,000 (assuming HR does not add in the extra costs for employees like insurance and so on) plus $2,400 for supplies and overhead. This gives us $74,800 yearly cost in years one through five, ignoring inflation. So, if you are not spending at least this much, or have very definite plans to do so, the purchase does not make economic sense.

If you are spending substantially less than the calculated yearly costs, then this is when using an outside vendor for X-ray imaging should be strongly considered. If you do have plans to spend this amount or slightly more, then a thoughtful discussion and business decision should be made. The pros and cons listed out and comparisons to outside vendors should be considered. Apart from the costs the factors of security, dealing with staff, available lab space, and needed certification by state and local authorities for some lab functions, as well as the time and effort spent by the company on work not in the core capability of the company should be considered.

If you are spending 2x the yearly cost of the set up you should very strongly consider doing this work in house. If the decision is made not to go with the internal lab you should then have a conversation with your outside lab(s) about the awesome deal they are going to make for you not to set up your own lab and continue using their services. Long term, a year or more, agreements with discounts or set costs should be used to secure what you need in terms of costs, turnaround times, and volume of samples.

It is important to have people in your company that can understand and interpret FA results whether the work is done internally or with outside vendors. Otherwise getting results from labs will be a wasted effort if no one in your company can understand what is possible in an FA or has any idea what the results mean then you will likely be paying too much for too little.

Chapter 2
Why is FA important?

Many people do not understand the benefit of reliability testing and failure analysis work. These people are short sighted. Every time there is a failed device, it is an opportunity to better understand the weak spots in design, materials, and production of the device and this can lead to improvements.

Anyone who complains that reliability and FA are just unneeded costs should think about the announcements on planes a few years ago that a particular brand of phone was not allowed on board due to fire hazards. The costs of recalls, lawsuits, and the missed opportunities to improve yields far outweigh the costs of FA and reliability. A line down situation can run into costs of millions of dollars a day.

Figure 2.1: Cell phone on fire. Image by Dan Sullivan.

I was on a plane several years ago when the announcement over the plane's speakers said, "If you have a XXX cell phone you will have to leave it at the gate or deplane." No one wants that publicity for their company and product. This was a direct result of management at XXX ignoring and/or skipping reliability and FA testing and results. When the product fails a qualification and the FA results show a problem, bad management will blame FA and try to get you to change or gloss over the issue to get the qualification to pass. When the disaster hits and massive returns occur, they will be just as quick to throw FA under the bus and state that, "It is your responsibility to report such issues."

You must never bow to pressure to give false or half-true results. It is far better for you to be fired because you would not fudge or shade the results than to knowingly state false results. Being fired for sticking to principles is a badge of honor to wear and will lead to other better opportunities as well as shield you from litigation

https://doi.org/10.1515/9781501524790-002

implications. It will also damage those who pressured you and that company will either have to replace them or fail. When I was told to stop doing reliability testing because some results were not favorable, I immediately started looking for another job. That company is of course dead now.

Bad things happen when FA and testing results in general are ignored. When the space shuttle Challenger blew up in flight, it killed all on board. I believe this tragedy is directly traceable to bad management ignoring materials and reliability engineers. Along with other issues, the O-rings used on the shuttle were made from a material, buna, that loses its sealing properties when cold. The tests showed this issue and other concerns for the heat shielding and the very poor management overrode these concerns and went ahead with the launch. This ignoring of data and results of testing led directly to this failure. If the O-rings had tests that resulted in failure and FA was done that showed what should have been the obvious, that the material used for the O-rings is not good in cold conditions and a different material or sealing system was used, then this tragedy may have been avoided.

Good companies do FA and successful companies learn from the analyses and improve. So, in the end, the failure analysis is a guiding light that enables us to make better/safer things.

Internal and external FA support is very important. If you are developing a new product or improving an older design, and you cannot inspect the results, you are not going to make much progress by trial and error without a feedback loop. If, really when, a defect makes the product fail for whatever reason, it is important to see and understand that defect and its cause(s) so the design or process can be improved. The reasons to want to do this are many, but the main ones are: the customer will not buy the product if it has an issue that is not explained, the customer will not continue buying a product if the field returns are going up and no reasonable corrective actions are put in place with evidence that they will address the issue, you do not want to make a crappy product out of pride, you do not want to injure or kill people out of a moral sense and/or fear of litigation and possible criminal problems, the warranty returns are going to end your business. All of these are legitimate reasons to want FA done right and quickly.

There are some industries where these issues are more relevant than others. If you make flip flops for people to wear around the pool you are unlikely to kill anyone, but you do not want chemicals from the flip flops to turn people's feet green. If you make a cat's toy and it breaks in a month no one cares, but it better last a month or the returns will get ugly. If you make IC chips that control a plane's navigation system you do not want any failures at all, as many lives and lots of money are in play and any incidents will be very high profile. Risk versus severity for failures will determine if you need to do an FA on each failure and what you are willing to spend on these efforts.

Ideally, through the testing and FA work, all the issues will be found, analyzed, and worked out before a product goes to market, or to any customers. This is rarely the case as time to market pressures are enormous and companies all strive to get to market first as market share is largely determined by who is there first. Business decisions can be very difficult and having accurate information available quickly is a great help.

FMEA (Failure Mode and Effects Analysis) is a process used to call out possible failure modes, their likelihood, and their effects. This enables testing to be prioritized to cover the most common or most severe issues. As issues are resolved, the FMEA is updated and testing priorities adjusted. While not directly FA, it often involves FA and drives testing. When used properly, it can save companies time and speed the product improvement cycle. When used only because it sounds fancy and no follow up occurs, it is a total waste of time.

Chapter 3
Planning out an FA

In approaching failure analysis work, it is always best to have a plan before touching a part. Initial optical imaging of the incoming part is also very important. Many steps in an FA cannot be undone and so an initial plan including review steps when results are obtained greatly increases the chances of success. Now, it must be said that FA is not a 100% success business. Often the process involves spraying hot acid on parts, grinding, and polishing to hit an exact or sometimes unknown location. Even the electrical checks can sometimes cause issues in the sample under study. One of the first questions should be: "Are there multiple samples or only one?" This answer will greatly affect the FA options.

When doing the initial steps on an FA with an electrically live part, there are precautions that should be taken. Many of these steps should also be observed on electrically dead or nonelectrical parts as well to avoid the inducement of false damage.

First ensure that ESD (electrostatic discharge) mitigation is in place, humidity is in acceptable range – usually 30–90% RH (relative humidity), floors are treated with dissipative wax or are ESD floors, and the people handling the samples are trained in ESD rules, wear smocks, shoes, and wrist and/or heel straps that are ESD compliant and tested every day. The samples are worked on table tops with ESD mats, the chairs are ESD type and linked to the floor (usually by a short chain), the parts are transported in ESD appropriate bags/boxes, and so on. A typical ESD workstation is shown in Figure 3.1.

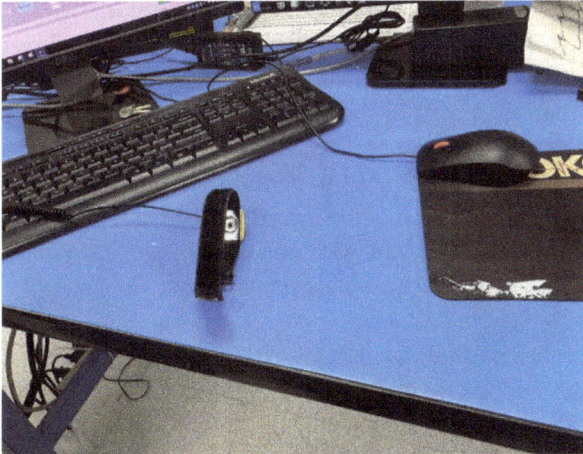

Figure 3.1: ESD work station for a C-SAM tool. Note the blue electrostatic dissipative mat and wrist strap near the keyboard. Image by Dan Sullivan.

https://doi.org/10.1515/9781501524790-003

There is no worse FA result than a false defect. It masks a real defect and has operations chasing their tails trying to solve an issue that does not exist.

I have received many samples that are not in proper containers and then customers wonder why the parts show a new failure signature at the lab. I have even had customers pull parts out of their pockets and hand them to me, complete with lint all over the surface. At least they do not shuffle their feet on carpet in the lobby and then touch the spot with their fingers to show where the issue is located. Oh wait, I have seen that too. Part of FA's job is to educate their customers.

If your company makes steel bolts, you will not be caring much about ESD but if you work with ICs and your company does not want to take the proper ESD precautions due to cost or difficulty, then you should be looking for a new job. You should also insist that the lab doing FA work for you has a robust ESD process including a SOP for handling ESD-sensitive samples. An example of typical ESD/EOS damage is shown in Figure 1.2.

If possible, you should have a written specification on how samples should be turned in to the lab. Share this with your customers as it is in both your and their best interest to find the actual cause of failures and not how carrying a part in a pocket causes damage to a sample. This is important for both general samples and ESD sensitive ones, and an additional reason to thoroughly optically image and document incoming parts. ESD bags/boxes are your friends. The lab may have to invest in a supply of these and provide them to the customers.

The first standard operating procedure (SOP) for a lab should be to have a specification and form for submitting FA requests. In this it is important to know when to be hard and when to let items slide. For a 5-min inspection with an optical scope or a check in the X-ray to see if the sample is correct or if anything can be seen at all, I suggest skipping the formal submission process. This will win you friends and keep paperwork that is longer to fill out than the actual analysis to a minimum. For these cursory, pro-bono inspections, it is a quick look only with no saved images. For anything that takes more than 15 min, or they require the images, always insist on the paperwork and the assignment of a job number in your system. In some situations, and with some customers, you might need to tie the release of the images and data, with the completion and submission of the form. If they do not have time to complete the submission form fully, then the work must not be very important and it should be put aside so the lab can work on more important issues. Do not let weasels run your lab.

Next week, when they need the images or a report for what was seen in a meeting and you are asked for the report, you will be glad you have a job number and everything saved in the right spot. They may try to get out of the formal submission process by saying it is just a little thing and the data stays with them. They lie and it will be your fault that there are no saved images later when your boss or theirs wants to see them. So, do not tolerate a loose work ethic for the data on jobs or submission processes. This will also be a great help when you are asked to justify current staffing or hires and new equipment purchases. If all the jobs are in the system, it is easy to later calculate the costs of the work if done at an outside service lab.

The idea of official FA process controls and documentation is often not well received by the lab staff. I used to think this way as well when I was young. I have learned the hard way why these documents are valuable and even helpful when used correctly. There should be an official document for submitting a job, which either has a form (traveler that will then accompany the samples) or a list of the information required for a request to be accepted. Without this, you are in the Wild West and you are not the cowboys. I have received samples with no instructions or just a part number. If you do not know who the requestor is and how to contact them then you are stuck. Just put such requests aside and forget them. They are obviously not very important to the customers as they have done such a terrible job submitting the work. If you have official documents describing the correct way to submit jobs, you are fairly bullet proof from this type of customer who just wants to dump the work and then blame FA.

It is important that you become an expert on the samples you and the lab are working on. It is not so important that you know everything right away, but you need to learn what the samples do and how they should behave. It is also very important that you learn the proper terminology. If you call the devices the proper name and all the components as well, it will not be noticed. However, if in a meeting or a report you call the solder bumps solder balls and the plugs M1 it will be apparent to others that you don't know the difference. The difference between solder balls and bumps is shown in Figure 3.2. This will adversely affect your standing with others and the trust they have in your results will suffer. It is better early on to admit a lack of knowledge and get the correct answers than three months into a project be caught using the wrong terms. This is even more important when dealing with paying customers. So have a pre-analysis meeting to get all the information you need from the customer including the proper names and terms. We saw some gunk on the thingy mebob is not going to go over well with the VP or the customer.

I have had to attend weekly meetings at one company just for damage control. Some engineers would hold onto work all week and then dump all the requests into the FA lab on Friday afternoon. In Monday morning meetings, they would then say, "I don't know, the requests are held up in the FA lab."

After a very interesting call with my boss and the VP of the packaging group it was decided I should attend the next meeting of the packaging group. Some of the engineers were not thrilled to see me at the Monday morning meeting.

Once I started attending these meetings and called out the day and time each request was submitted, this behavior stopped. Unfortunately, it required me to attend meetings in a different building from the lab in person to put an end to what should have been an obviously wrong approach. It is always better to sacrifice some time up front to correct a process gone wrong than to wait and hope it fixes

Figure 3.2: 3D X-ray image of FC device with solder bumps (little ones) and solder balls. Image by EAG.

itself. This may mean an unpleasant discussion or confrontation, but better early than once it reaches unbearable levels.

Next, you need written procedures for each of the major activities the lab performs: Receiving work, optical inspection, decap, SEM, and so on. These procedures should not go into painful detail. For example, in a cross-section SOP, you do not need to call out each type of polishing sheet to use and for how long, just write "use polishing sheets as appropriate and until scratches are removed or all scratches face in the same direction for each polishing sheet used." The training should then fill in the gaps. These are also very useful as cheat sheets for techniques the staff do not use every day. A section on how to turn on the tool, collect and save a spectrum and/ or image and a section on how to properly leave the instrument is very helpful. Also, to be prepared for unfortunate events when the power goes off suddenly or the machine needs to immediately be shut down, a section on how to properly power off tools and how to restart them is great. The contact information for service and vendors for the consumables or frequently replaced parts should be included in this document and all the lab staff should be trained on this process.

I almost hate to have to say it, but these documents are worthless if you do not use them. Train people with them. The first step to learning to use a tool should be to read the SOP (standard operating procedure) document for the tool. The SOPs must be either hardcopy in the lab or *easily* accessible online. Talk with your QA people to get this right and insist they do it the way you want – easy to access, or do not bother. I once had QA tell me they were the only ones that could have these documents online and it took only nine layers of nonsense to get to them. Guess how many times these were accessed by lab staff. These geniuses did eventually let me have hard copies chained to each tool as long as they were laminated and had the rev number and my signature on them. Sometimes, bureaucrats get confused thinking the rules are the reason instead of the actual work.

While I do understand that in a factory you want to ensure that the latest specifications are being used and there needs to be some control over this it needs to actually work. Making it very difficult to use the system just leads to people ignoring the system. Being able to shrug and say "Well they did not follow our procedure" at the end of a disaster is not very helpful. You may need to get support from upper management to make QA do their jobs in a helpful way instead of what is easiest for them.

Chapter 4
Typical types of failure mechanisms

Of course, the typical failure mechanisms will depend on the type of samples you have and their history. Here we will cover those we have commonly seen in the IC and electronics industry. Starting with the wafer or die level failure modes, the most common are: physical handing issues, leakage, high resistance, and opens/shorts.

Physical damage is usually caused by dicing, pick-and-place tools, or humans (especially with metal tweezers). There are a variety of dicing methods in use and they vary in their rates of damage and the way they cause damage. Figure 4.1 shows cross sections of several dicing methods. Usually, the dicing damage is caused by a system that is not well maintained or has not been used on this new device. I have seen edge damage from dull blades, loose "chattering" blades, and blades that hit objects in the street (this was a truly dumb idea to place metal critical dimension (CD) markers in the street). Typical scallop damage is shown in Figure 4.2.

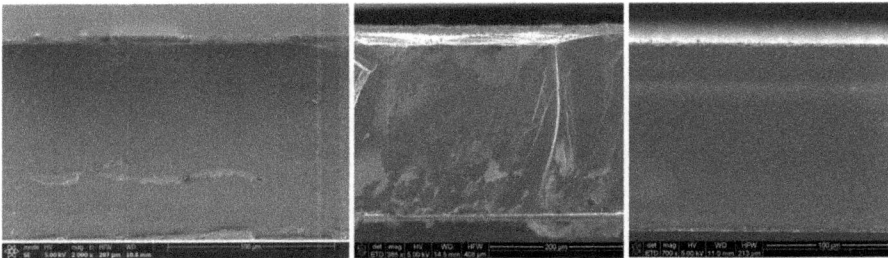

Figure 4.1: Die edge damage on two die (horizontal and vertical damage) and a new method of dicing that leaves very little damage. Image by EAG.

Pick and place tools can hit the edge of die if not properly aligned and the rubber suction cups do wear through eventually and cause telltale circular or partial circle damage on the die when the metal tube behind the suction cup is exposed and contacts the surface. This is due to poor maintenance or assembly set up. This type of issue requires inspection in the assembly process after each step to catch and identify the tool(s) responsible.

Package damage is more common before a device is attached to a PCB. The corner damage shown in Figure 4.3 regularly results from a dropped part or when devices are improperly seated in a Jedec tray and the lid forced closed.

Humans can cause all kinds of damage, but the most common is the metal tweezer scratches. One issue I witnessed involved assembly of a die which had very good yield on one lot and the next lot had very poor yield. This occurred several times with high yields followed by very poor yields and the FA team observed large

https://doi.org/10.1515/9781501524790-004

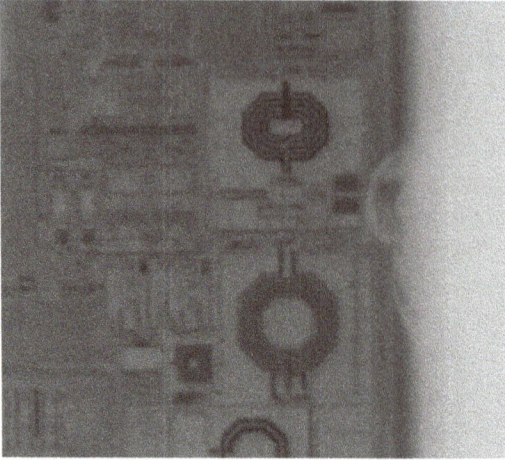

Figure 4.2: IR image of scallop damage. The damage is seen through the ~700 microns of Si. Image by EAG.

Figure 4.3: Package damage. Image by EAG.

scratches on the face of the die for the failing units. This cracked the passivation and led to failures in stress testing. When we had observers put on the assembly line, we found the metal tweezers being used to handle the die by one shift were causing the scratches such as those shown in Figure 4.4. Issue solved right? No, the issue came back the week after we removed the observers. We had to have observers go back in for several weeks and retrain the assembly personnel and confiscate some tweezers. Some issues are tougher than others to resolve.

In the case just described, the cause of the passivation cracking issue is obvious. This is not always the case and it is important to consider all possible sources

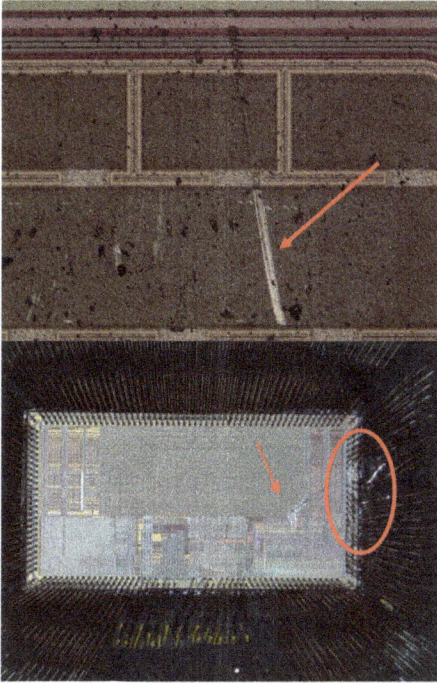

Figure 4.4: Die face scratches from tweezes, and typical die scratch and wire sweep (inside oval). Image by EAG.

for such damage including inadvertent introduction of scratches and passivation cracking from the FA process and improper electrical testing.

Leakage issue can arise from many root causes. The most common I have seen are: particles and poor etch process. Particles are shed onto the wafer during a variety of deposition and etch steps in the Fab. One of the core results of interest in such an FA is what is the particle made of and the other is what layer is it on. These two items help identify what chamber/process has the issue and needs to be serviced.

The deposition and etch steps of making a wafer involve placing resists down, patterning them, removing the desired material, depositing the metal, oxide, nitride layer and then etching away the extra material and the remaining resist. If the resist does not do what is intended or the etch process is not correct (under or over etch) then too much or too little insulator/conductor may be the result, which leads to leakage paths, opens, or shorts in the device.

The issues that cause leakage are often the same ones that generate opens and shorts and high resistance. In these cases, a larger particle or in a less fortunate position or alternatively the under or over etch is more severe creates a more severe failure.

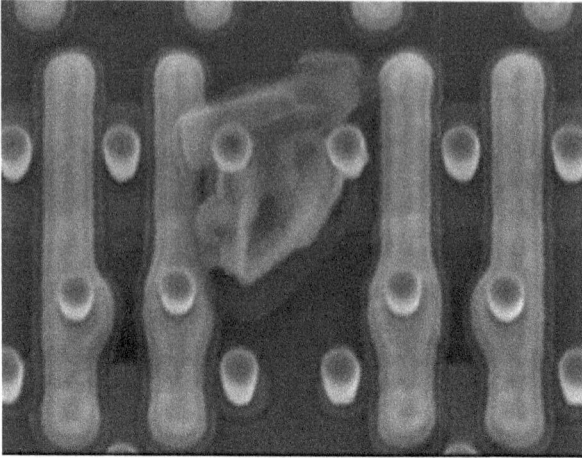

Figure 4.5: Image of particle in IC at the W plug layer. Image by EAG.

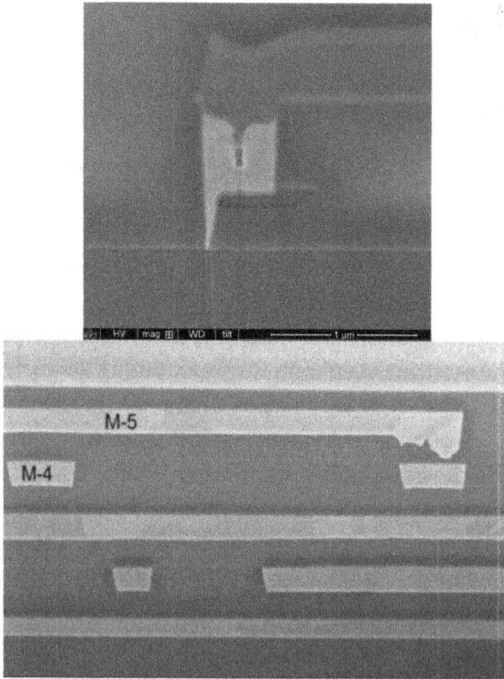

Figure 4.6: High-resistance contact and M5 connection failure. Image by EAG.

For packaged devices, the above issues in the die may occur on the die inside, but the issues attributable to the package itself most commonly seen involve: solder, delamination, and wire bonding issues.

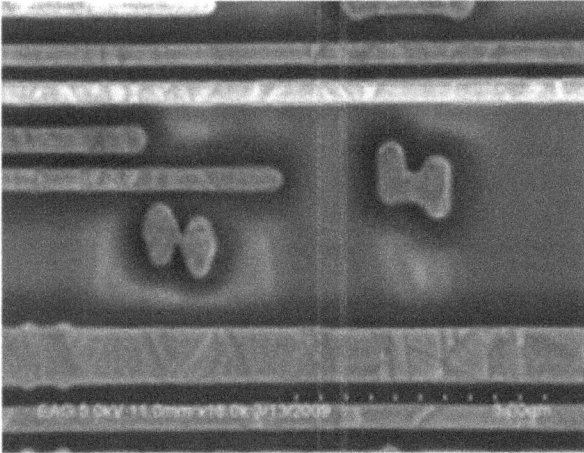

Figure 4.7: Under-etch pattern leaves too much contact in place. Note: this can also be caused by a patterning issue and the reticle should also be checked. This failure is sometimes called "kissing" as the adjacent metal are supposed to be completely separate but like Romeo and Juliet they find a way. Image by EAG.

Figure 4.8: Cross section of solder non-wetting, so-called head in pillow (HIP) or snowman bond. Image by EAG.

A common solder issue is non-wetting. This occurs when the solder does not reflow due to an incorrect reflow temperature profile or contamination preventing the solder from making proper contact. Your assembly house should always have the reflow temperature profile for every run. If they do not or will not share it, find a different assembly house. The issue of contamination is harder to get to root cause as the part has been through a reflow which will volatilize many common organic contaminants and getting to the surface of the solder after assembly is not straight forward. This will require the assembly house cooperation to check for contamination at several points in the assembly process and involve tools like FTIR, XPS, Auger, and EDX.

The typical form of this is called head in pillow (HIP) or snowman non-bond. HIP issues can regularly be found by X-ray imaging (2 and 3D imaging) or by a cross section like the one above in Figure 4.8.

Figure 4.9: FC delamination in the die. This case occurs in a low K dielectric layer. Image by EAG.

Delamination is the separation of layers in the device. These include: the encapsulant to the die surface, layer(s) inside the die (I have only ever seen this with low K dielectric layers), the die-to-die attach, and for flip chip devices the underfill to die or underfill to substrate interfaces. These separations cause electrical opens by pulling the wire bonds up or disconnecting conductive die attach from the backside of the die, and in flip chip, the solder bumps can crack as the underfill no longer supports them. Even layers in the substrate can delaminate. Scanning acoustic microscopy C mode (C-SAM) is the common technique for detecting these issues.

In Figure 4.9, the bright area in the lower right corner is the delamination. This is likely in the low K dielectric layer in the die, but a cross section should be done to verify.

Wire bonds come in many flavors. The standard wire bond materials are: gold, aluminum, and copper. I keep hearing that silver is just about to take off, but I have heard that for five years now and I have yet to see one silver wire bond part except the one we had to be begged to get so we could test decapping it. A typical wire bonding failure is shown in Figure 4.10. The wire bonds can have several attach styles as well. The two main methods involve ball and wedge (also called stitch).

Figure 4.10: Lifted ball bond from inadequate wire bonding. Image by EAG.

These bonds can be double layered and various other combinations including the addition of other materials to help make the bond tougher. I have seen a stitch bond with a ball over it. I have seen silver paint under a stitch bond. These all seem idiotic to me as you now have several interfaces that can fail instead of one and you have applied multiple stresses on the bonding. I have yet to see a study comparing these methods as most of this information is kept as trade secrets by companies and so not made public.

The wire bonding process can create problems. First and most amusing from the FA point of view is when the wire bonding program is incorrect or the device orientation is incorrect during bonding. This is easy to catch as the X-ray image will show the wires not going where they are supposed to in the layout diagrams. There is a pin one location on all devices that I have observed that sets the device orientation. The X-ray image is compared directly to the layout and if the wires do not line up, that is the issue. FA done and report goes out, easy. I have even done a couple of FAs when the bonding wires, once even the die itself, were missing. Now someone in the assembly house has some explaining to do.

The wire bond process involves pressing a wire onto the bonding pad or surface, commonly Al or Au pads, with force, heat, and ultrasonics. This process is worked out by the assembly house, but occasionally they get it wrong. This can be because the materials are new, or the wrong program got loaded or it is not a very good assembly house. The "wrong" can result in either overstress which breaks or craters the die, or understress which results in nonadhesion, and sometimes the

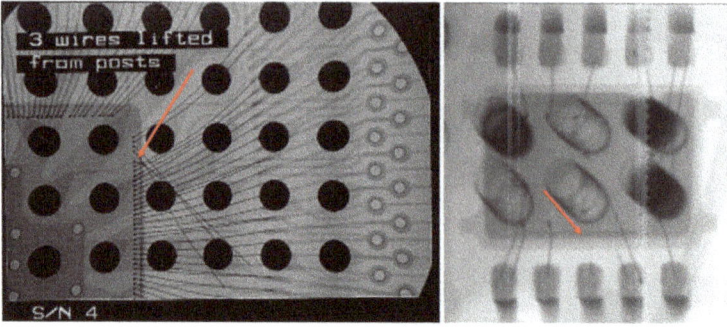

Figure 4.11: X-ray image on the left showing several wires lifted from posts on the substrate. The image on the right shows bond wires that have been melted. Image by EAG.

wrong intermetallic phases of the wire and pad are created, like the purple plague. The purple plague is named such as it has a purple color and is very bad as the phase created in the AuAl interface is very brittle and crumbles causing opens. This issue of purple plague was understood and avoided in the 1990s but occasionally creeps back in at some assembly houses.

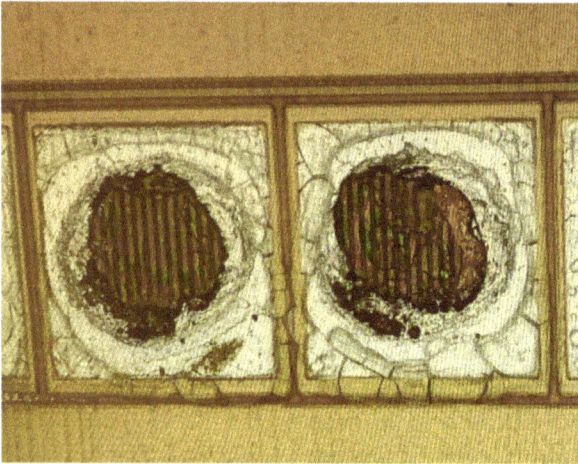

Figure 4.12: A cratered bond pad. If the force and/or sonics used are too high, then the bonding will fracture the underlying die and the layers of oxide and traces and even down into the silicon will crack and pull out when the part is decapped. Image by EAG.

The high-end wire bonding tools are amazing to watch as they make many connections at high speed. They make me think of giant high speed sewing machines. When these have alignment issues, it is usually obvious quickly and gets caught by QA inspections in line before the parts get fully assembled. For wire bonding on

smaller scale, this is not the case and issues do wind up in FA. Occasionally, mis-
alignment is so bad the ball bond is not well centered on the pad and this can crack
the passivation around the pad leading to leakage issues. The ball bonds ideally are
dead center on the pads. The further away from this the more likely issues will arise.

Figure 4.13: Slightly off-centered bonds may not affect the part performance, but once the edge of
the pad is hit and the passivation surrounding the pad cracked this will lead to leakage, shorting,
and reliability issues. Image by EAG.

The standard wire bond lift FA involves an electrical check at the bench to verify the
open/high resistance. Pushing down on the top of the device may cause it to recover,
and this is a common sign for ball lift or solder issues. The X-ray analysis will show
gross lift but very small, less than 3 microns, is difficult to observe. C-SAM is then
done to check for delamination in the wire attach areas which will lift the ball or
stitch bonds. The destructive work then begins with a decap. At this point, the wire
will lift or very easily be moved by light pressure as the encapsulant is no longer
holding it in place. The bottom of the bond and the area under the bond are now
inspected by optical microscopy, Auger, and SEM to check for contamination, thick
oxide or excessive intermetallic layer (IML) growth. The common results are as fol-
lows. No bond was ever made or the pad is cratered both result from incorrect settings on the force of the bonder. The oxide on the pad is very thick and a bond was
made but did not adhere. Gross contamination is present and visible. The surface is
corroded and the IML is thick and crumbly (check for Cl contamination).

Purple plague is a special case of IML issue. This occurs when the conditions
during the bond creation are incorrect and undesired phases of the intermetallic
are made. The one called purple plague is very brittle and leads to cracking. It is
also a very pretty shade of purple.

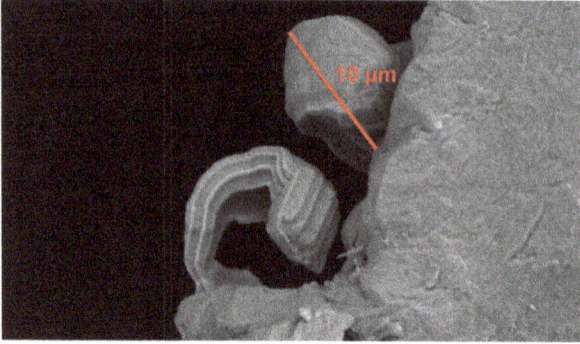

Figure 4.14: Tin whiskers grow out of solder and can get quite long causing leakage and shorting issues. Image by EAG.

Tin whiskers can grow out of tin containing solder, an example of which is shown in Figure 4.14. Usually, strain is the driving force and the whiskers can grow quite large and cause shorting or leakage issues in a device. A large amount of effort has been made to understand and eliminate the conditions for whisker growth.

Dendritic growth can occur with silver or copper under the right conditions. I have only ever seen dendrite from field returns and biased damp heat reliability testings. The potential directs the growth and the humidity works with some ions (typically Cl) to etch one copper trace or the silver epoxy die attach and moves the metal toward the nearest ground or negative bias spot. This grows dendrites or halos which will eventually result in leakage and shorts. An example of dendritic growth in shown in Figure 4.15.

Corrosion can happen when the etchants used in the FAB process are not completely removed (typically Cl and F). An example of a corrosion is shown in Figure 4.16. These can corrode the Al pad and the IML. This is easy to see once the part is decapped, but finding the culprit (Cl, F, or another etchant) is difficult as the decap process will wash them away. This requires a dry-etch approach which typically takes longer and requires a higher level of skill.

Truly bad or unlucky assembly houses also create issues by not properly cleaning the die and package components prior to assembly and the contamination present will cause opens and high resistance by interfering in the bonding process. Also, faulty tools or clumsy operators can cause wire sweep and allow stray material (typically wires or solder) to fall onto the die surface leading to shorts. Occasionally even finger prints have been found on die which are excellent sources for Na, K, and Cl all of which do bad things to devices. Having a good assembly house is a very big deal.

One assembly house I dealt with had developed a cleaning process involving plasma on the die and the copper lead frames. They did all the work with a single lead frame at a time. When they had the process set, they the loading 120 lead frames and were surprised to get many bonding failures. It turns out that the lead

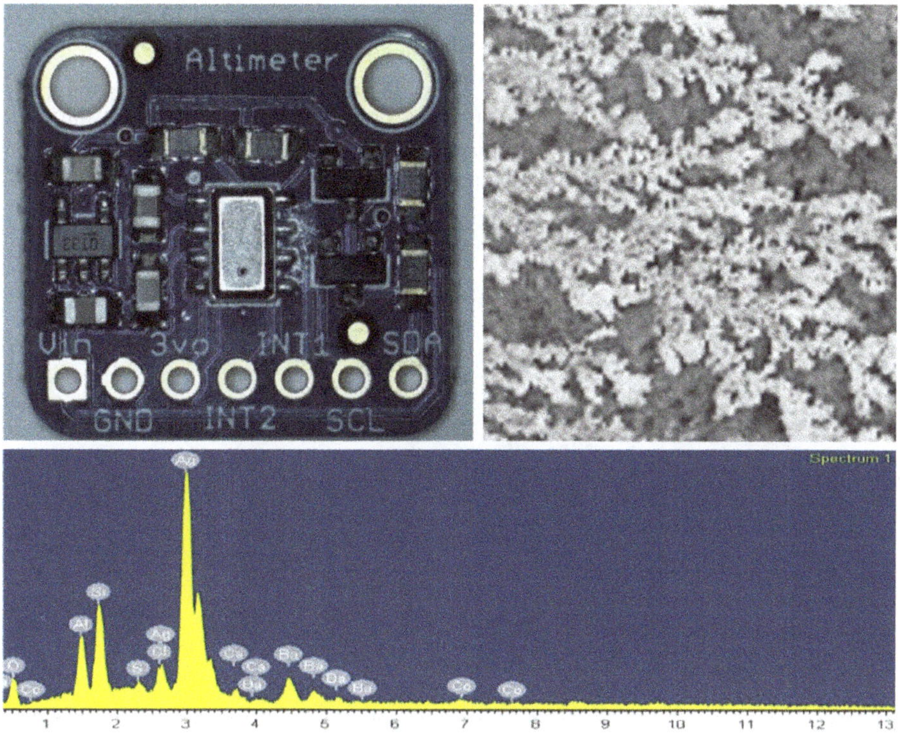

Figure 4.15: Ag dendrites on a sensor board. EDX identifies the dendrites as Ag. Image by EAG.

Figure 4.16: Corrosion on bond pads. The brown areas in the bond pad squares are where the Al has been etched away. Image by EAG.

frames in the center of the plasma tool were shielded very nicely by the surrounding lead frames, so about 40% were not cleaned at all.

Of course, there are many other defects found, but we cannot cover them all in just one book. Each year there is a very good conference, known as International Symposium on Test and Failure Analysis (ISTFA), on FA that has several papers covering examples of FA work on specific types of defects.

Chapter 5
Destructive versus nondestructive analysis

In approaching failure analysis work, it is always best to have a plan before touching a part. The steps of the plan are followed and choices made after each result is obtained. The problem with a locked in plan for FA is that after each step, the results determine what the best next step is and sometimes there are several options that make sense. A plan with multiple branches and some question marks is what we typically have in FA.

The first steps should always be the nondestructive analyses and verification tests for the suspected failure mode. These leave the sample basically unchanged. Nondestructive testing (NDT) is not always completely without effect on the parts but is usually so. The most common exceptions are that X-rays can affect very vulnerable memory; in the DI water, the part which must be submerged in for C-SAM can cause electrical leakage and some materials that are water soluble may be removed. But the most common issues are with handling: dropped samples can be damaged, ESD events can occur if the proper precautions are not taken, and even improper placement in carry devices can damage the samples. This is part of why it is always desirable to have multiple samples, but this is not always possible.

The order of analysis may be changed depending on the exact failure mode, but normally proceeds in this order: Optical inspection, electrical verification of the failure mode, X-ray to see if any wire bond or solder issues are obvious, and then C-SAM to check for delamination on the die or wire bond area, or in the case of flip chip devices delamination in the die or at the underfill interfaces. An example of a failure observed in C-SAM is show in Figure 5.1.

The electrical verification of the failure mode is very important as the parts can recover or incorrect parts may have been submitted. If the failure mode is gone it is very difficult to find the root cause of an issue that is not there. You do not want to spend your valuable time chasing nonsense. If the wrong parts were sent, then the right ones can be sent. If the part(s) recover that actually points to likely issues. Leakage due to water in the part or dendritic growth in the parts can both recover over time or with a bake. Many labs actually do a quick 12 h bake at 100 °C before continuing on with an FA to ensure they are not chasing these failure modes in the lab. The dendritic growth can sometimes be caught by p-lapping or plasma removal of the encapsulant to reveal the die without removing the delicate dendritic growth. If you know that you are looking for dendrites you have a chance but with water ingression you never find directly only through indirect evidence. Placing the recovered part in a damp heat chamber or submerging the device in water for at least 4 h to check if the failure mode returns for instance.

Other less common NDT methods used in FA include: time domain reflectometry (TDR), IR (infrared) imaging, and FTIR (Fourier transform infrared Spectroscopy).

https://doi.org/10.1515/9781501524790-005

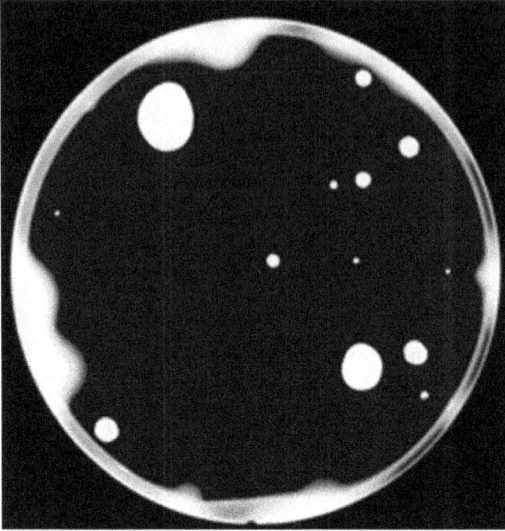

Figure 5.1: Reflection mode C-SAM image of bonded wafers with air gaps between the wafers (white areas). Image by Lisa Logan.

TDR uses a probe setup to contact the pin or solder ball in question and an adjacent ground. A pulse of electrons is sent into the sample and the reflected signal is captured. At each interface in the path of the electron pulse, a portion of the signal is reflected. So, the solder ball to the pad, each interface in the bond pad, the vias connecting each trace to the next level in the substrate (top and bottom of each via), onwards to the bond wire (or solder bump for flip chips), and then into the die. So, usually about 10–30 interfaces. This produces a squiggly line for a device probed at a particular input. Some people claim they can tell what each turn in the line indicates and its location in the device. I find this hard to believe. Fortunately, this special power is not required to understand the analysis.

Using the failed unit (FU), a known good part (KG), and a good substrate (GS), a comparison can be made and the location in the spectrum that the failed part deviates from the KG or GS signal provides the general location of the defect. If it is before the GS overlap with the KG then the defect is in the substrate. If the deviation occurs just at or just after GS then the defect is at the interconnect (bond wire or solder bump) and if the deviation is further in, then the issue is in the die. Now if the signal does not deviate from the KG then breathe and recheck your setup as you are either testing the wrong part, the wrong pin, or the part has recovered.

The preparation of a GS if one is not available from the customer is straight forward if you have an additional KG device. Simply grind/polish the part to break the bond wires or remove the die for flip chip devices. This should be a quick 10-min operation at most.

TDR is what I call the unparalyzing tool. I have experienced a meeting in which discussion of what to do for the only failing part in a qualification for a major customer took the following horrible path. The device fails and does not recover. There is no obvious physical damage. We need to show root cause of failure or the customer will walk away. Ok, can we decap to examine the die – no the failure might be on the die surface. Ok, can we remove the substrate and examine the die through backside analyses – no, the issue could be in the substrate. Ok, can we break the bond wires and electrically check both sides – no, the issue could be between the bond wires. So, we can just look at the device intently and wait. We are paralyzed.

The TDR makes this problem go away and we can identify the region of the issue and start to do the destructive FA. This also lets people know which, the substrate, assembly, and die, has the issue and they can go and starting examining the issue in the right place.

IR imaging is used to look through silicon to examine the circuitry in a die. Typical use is to examine a flip chip device for defects and cracks, especially at the corners or scallops along the edges of the die. The technique is fairly quick but is usually on an expensive instrument. The defects found in this way commonly result from dicing operations. Normally there is some roughness on the sides of the die after dicing, but any damage that extends past the end of the street and into the circuitry is an issue. The damage may cause immediate failure or create a reliability issue as the cracks may increase in size over time.

Fourier transform infrared spectroscopy (FTIR) is a tool used to identify organic materials [1]. When a sample in an FA has a grease or oil or stain that appears to be organic in nature, then FTIR can be employed to identify the material. FTIR instruments come in two main varieties: low cost with a large sample area needed or one with a microscope which can look at a small spot of ~15 microns. The IR does need a healthy, 50-nm-thick analysis depth. So, very thin contamination areas cannot be analyzed. The use of attenuated total reflection (ATR) can make the analysis more surface sensitive as the IR beam is passed through the surface layer many times to increase the absorption by the contaminant. The spectra collected from the FTIR work should have a blank spectrum subtracted from it to remove the peaks present from the air.

To determine what materials are present, the spectra can be examined and each peak or set of peaks can be assigned to a particular functional group: OH, CH_2, and so on. An easier way is to use libraries of spectra that are commercially available to match to the acquired spectrum. There are large libraries of spectra from known materials with hundreds of thousands of spectra available, and of course the labs can add any spectra they have done themselves. In this way, very quickly, a best match can be obtained for the unknown contamination.

A good practice in a fab support FA lab is to collect all the known materials being used in the fab and create a spectrum for each to be used in their library searches. If the bandwidth is available, the materials can also be exposed to heat and UV to get the spectra for the degraded versions as well.

There are many other materials analysis techniques that can be used in FA work, but these are less common. Doing some research in the literature or contacting a service lab will help determine the best course of action when these less common techniques are required.

Chapter 6
Optical Microscopy (OM)

Typically, every FA group has at the very least an optical microscope. Most will have two with a camera for capturing the images. The two scopes are usually a low magnification scope, 20–200x, which is trinocular, two tubes for your eyes and a third above for the camera mount. The higher magnification microscopes have many variants; from the low-cost versions with five or so optical choices ranging from 5 to 150x and a 2–20x increase in magnification for the rest of the system. So, the final magnifications are generally 10 to 3,000x. The highest magnification lenses often require immersion and I avoid these as they are messy, have very short focal lengths, and can introduce surface contamination from the immersion fluid. Just use the SEM at that point.

There are straight bright field microscopes, although the high magnification scope can have dark field optics as an option for some of the magnifications, this is usually a decision made based on cost. Having one dark field optic is a great option for those cases when bright field does not provide the right contrast. The low-cost "Dan Sullivan" version (I was told of this package by a vendor after I had bought this set up for the third time): A low magnification scope (20–150x) and a higher magnification scope with five objectives of which only the 10x optic was bright and dark field. These two scopes share a single 3 MP camera and one computer with capture and image processing software.

If money is not an issue, then many alternative optical microscope varieties are available for FA. Many scopes now have bright field and dark field along with mixing and angle options. These enable shading and very surface-sensitive images to be obtained. A motorized stage is very helpful for calibrated measurements as well as enabling multiple images to be taken across the sample which are stitched together by the software to make a large area image at maximum magnification and resolution. This is a great time saver and stops the endless requests for "just one more image here at higher resolution." Large die, 16 mm square or larger, can be imaged at maximum magnification and resolution and presented as one image. If the customer wants to zoom in on an area they can and the resolution is already the best it can be on that system. These scopes are about three to four times the cost of the special.

Also available at the higher end are confocal microscopes which have very narrow focal planes so different layers in a sample can be observed. This in theory removes the need to then delayer the samples to determine which layer the issue is on. I have yet to have a customer accept this and have always had to do the follow up delayering or cross section to image the defect directly.

Additionally, at the high end are the infinite focus scopes that allow a continuous changing of the magnification instead of the step magnification provided by

https://doi.org/10.1515/9781501524790-006

multiple optics scopes. These are very nice in that the area of interest can be zoomed in to include whatever is desired instead of a shot at 5x in which the defect is very small and then the next step, 10x, magnification in which the defect more than fills the screen.

Some systems have autofocus features which can be used to get all layers in the images to be in focus in the captured image. The same feature can provide Z measurements and easily determine if features are boulders or divots which can be difficult to determine from just a top-down image.

Shown in Figure 6.1 are images of PCB damage with bright field, dark field, and dark field with the sample covered in IPA. The dark field with IPA allows clearer imaging of the buried metal trace. One of the advantages of optical imaging is that some materials can be seen through as is obvious in the images of the PCB. The limiting factor of optical microscopy is the magnification/resolution possible. Without considering laser imaging, the maximum magnification is ~2,500x. With immersion lenses and some tricks, it can be slightly higher. This is good, but often insufficient for FA on modern electronics/semiconductor devices.

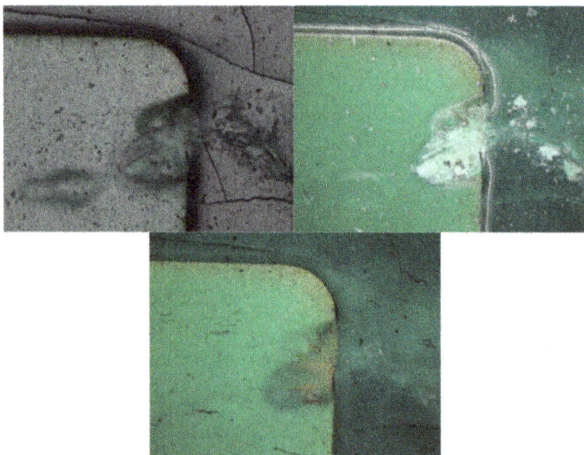

Figure 6.1: Optical images of PCB damage with bright field, dark field, dry and covered with isopropyl alcohol (IPA). The IPA fills in the scratches on the surface and allows clearer imaging of the buried layers. Image by EAG.

The use of optical images for documenting samples and inspections should be universal. If you set up a general FA SOP, it should always have step 1: Take optical images of the sample(s). If there are multiple samples, then take an image of a normal sample and any deviant devices as well.

Unfortunately, many labs use the optical microscope for initial inspection, but do not save the images. This is counter-productive as these images make it much easier to understand a report, visualize what the sample is, the locations of analysis/defects, capture the condition of the sample when received, and after certain operations are performed on the sample. These often seem obvious to the direct customer and the analyst, but a few weeks later when the VP of the major customer wants an explanation in detail these images are very helpful. Explaining that the samples looked worn and not quite right is helpful, but an image showing the condition will speak for itself. Electronic images are free, take many. A great many times I have received a call after completing an FA with a request for the date code or part number of the device. These are almost always found on the device itself, but after an FA the device has been decapped or worse and the printed information is no longer present. A couple of images of the as-received part(s) solves all these requests, such as the image in FIgure 6.2.

Figure 6.2: As-received device on a PCB. Note that the information on the device in the image allows for retention of information will be lost during the FA process. Image by Dan Sullivan.

In some cases, the optical images may be all that is needed to wrap up an FA. The missing solder ball, the huge crack in the package, and the fact that the part is completely burnt or not the right device all close the FA book. Large EOS damage captured by the OM such as the example shown in Figure 6.3, also falls on the done list.

A special note here on naming data files. This will be true for all data files and I mention it here as Optical images are generally the first collected.

For the naming of files, any scheme you wish to use that is consistent should work. I have adopted the following when possible: the job number is month (03), day (15), and year (2022) for the ides of March this year and then a number 01. So, the third job to come in today would be 0315202203. This is the name of the top folder for this FA. The images I then separate into folders for each type of work (optical, SEM,

Figure 6.3: Optical image of some Electrical Overstress (EOS) damage in a die. Image by EAG.

etc.) The images get named with the job number, sample ID, and then a 1, 2, 3, and so on. This seems like a lot of work, but it makes finding items later much easier, especially for someone other than the analyst that did the work. You will be much happier with a good system when you have to answer a customer about work from two months ago done by Johnny who is on vacation.

Chapter 7
Scanning Acoustic Microscopy/Tomography (SAM or SAT)

This nondestructive technique uses sound waves to scan a sample. The reflected and transmitted signals are captured and images of the interfaces selected are created. The samples need to be submersed in a liquid medium for analysis. Normally, DI water is used. Other liquids may be used like IPA but the smell, cost, and evaporation rates make water the most often used. C-SAM instruments are essentially fancy air bubble detectors. The speed of sound is very different in liquids and solids compared to air. This large difference results in a very high reflection percentage at air interfaces. Thus, air gaps result in very bright areas in reflection images and very dark ones in transmission or thru scan images. Samples must be fairly flat with surfaces that are not too rough; otherwise, the sound waves will be reflected away from the detector and no signal will be observed.

The resolution in X and Y dimensions depends on the material scanned and the transducer frequency but is typically no better than 15 microns, while in the Z dimension it is very thin, ~0.01 micron air gaps can be detected. The higher the frequency of the transducer, the better the XY resolution. However, as the higher frequency signal is more readily absorbed by materials, this reduces the depth of penetration into the sample. So, thick samples are not as readily analyzed by the higher frequencies.

The transducers are made with different focal points and this also affects the thickness of samples and the width of the focal plane (Z resolution). Since each transducer is between $500 and $12,000 USD today, having a complete set can get very expensive. Typical FA sets include 10, 20, 30, 50, 100, and 150 (230) MHz. Different instrument manufacturers use slightly different numbers but this set is common. For examination of smaller devices, a few mm square, a 75 or 90 MHz are often added.

To examine typical IC packaged devices, 30 and 50 MHZ are the most commonly used. For flip chip (FC), the heat spreader is attached to the substrate by adhesive and by thermal grease so the die can be analyzed as received with the same transducers. But if the substrate/underfill and die/underfill interfaces or internal layers in the die itself are of interest then the heat spreader needs to be removed and any adhesive on the die cleaned off. The 100 or 150 (230) MHz transducers are used when analyzing the FC die and underfill interfaces depending on the thickness of the die.

Removal of the heat spreader varies from very easy using a razor blade or modified wafer tweezer in a few minutes to soaking overnight in solvent, or polishing/chemically etching away the heat spreader. Using the cut and pry method on the tougher samples usually results in cracking and shattering the die. The polishing

https://doi.org/10.1515/9781501524790-007

method always works but takes the longest and requires some training/skill. The heat spreader is not completely removed by polishing but instead when it is thin enough, it can be peeled off. This method ensures no accidental grinding on the die occurs. Once the heat spreader is removed, there is almost always some greyish material on the die. This is either thermal grease or a thermal epoxy of some sort. Its function is the transfer of heat from the die to the heat spreader and lends some structural integrity to the device. The thermal grease is usually easily removed by a cotton swab with IPA and elbow grease.

Sometimes, the thermal grease is dried out or the thermal epoxy is tough. This requires some force and scrapping of the material to remove it. Do not use metal tweezers to do this removal. Best case is when you are so skilled it works, but usually it will crack or badly scratch the die. Use a wooden tool, a broken cotton swab stick works well. The wood is not hard enough to damage the die and will remove the material with some work and IPA. If the material is not removed it will show up in the C-SAM images as it absorbs or scatters some of the signal. Figure 7.1 shows optical images of devices at varying degrees of material removed.

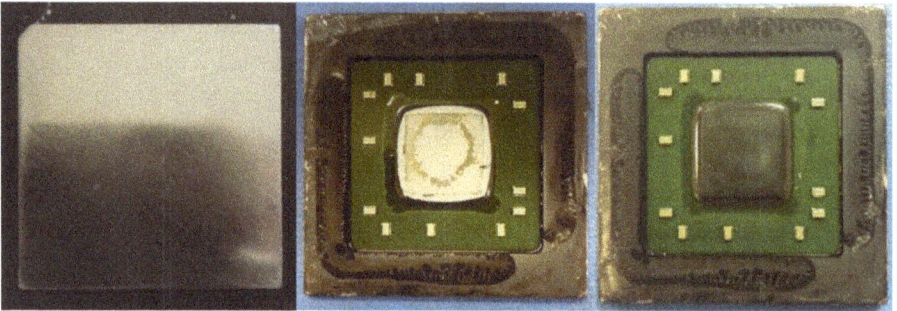

Figure 7.1: Optical images of a FC device with heat spreader (HS), HS removed, and thermal attach cleaned off. Image by EAG.

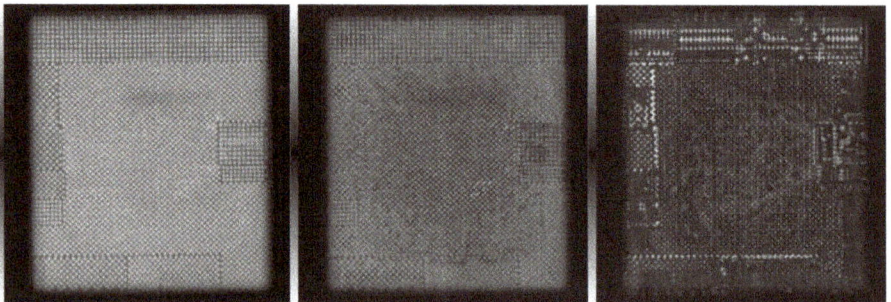

Figure 7.2: C-SAM images of FC device with set Z, scanned with images at three depths: Die to bumps, bumps alone, and bumps to substrate. Image by EAG.

The most common use for the C-SAM is in reliability studies to examine all the samples prior to stress and after stress, sometimes at intermediate points along the test timeline. JEDEC Specification 883 calls out the pass/fail criteria used. All reliability testing uses of C-SAM on ICs follow these guidelines when looking for failures. Delamination may not be present on the die surface or any wire bonding surface. Delamination may not grow during the testing or go from one edge of the device to another edge.

In FA, these same guidelines generally apply. If there is a suspected delamination observed in an active area that is always bad; in other areas it depends. C-SAM is usually not the last step in any FA. It is more of a technique to call out that something is wrong and should be further probed. It may be acceptable to call out delamination directly from C-SAM images. Before an anomalous area on a C-SAM image is designated as delamination, there should be a solid history of cross section confirmed cases on the same type of part. An example of such delamination is shown in Figure 7.3.

Figure 7.3: C-SAM image of underfill delaminating in a Flip Chip device. Image by EAG.

This is a technique that needs an experienced and honest operator. It is fairly easy to make good samples look like total failures and vice versa. Samples with some delamination, but not total delamination, are harder to disguise, but it can be and I have seen it done.

I once worked at a company where I took over a very dysfunctional lab. In reviewing the older data, I asked if all the old data can be thrown away or do I need to start calling past customers to inform them that their passing parts really failed. In these old jobs, the focus was not right and the contrast was lowered along with the gain so everything was washed out, but I could still see the delamination areas. The technicians, both of whom were gone within 30 days of my arrival, did not understand the physics of the technique and just tried to make the images match older images in any way they could. Part of this was fear of results showing failure and part was because they did not understand what they were doing and the repercussions of what they were doing. Trained staff that are encouraged to always tell the truth no matter the consequence is very important to any lab, but especially in FA where unpopular results occur and the staff have to be aligned to the truth and no other line of thinking or they cannot be trusted, especially with techniques like C-SAM where the result can be faked by very skilled or very unskilled operators.

I have been asked many times by people who have sent me C-SAM data they got elsewhere if the delamination they see in red is real or not. If they send me only the reflected scan image with the red/yellow/gray scale, I tell them they have a very nice picture that their child could have drawn for them, but it means nothing. The reflected scan only, especially the red/yellow/grey one is useless unless you know the source and trust it. You really want the Thru scan and a couple of A scans (in a good versus delamination area) to verify the results. If your FA provider is unwilling to provide a couple of A scans then you should run from them as fast as you can. They either do not know what they are doing or worse they do and are not to be trusted.

Figure 7.4: C-SAM results for the top surface of the device, die surface scan, and then Thru scan with air trapped between the device and the PCB, with the air bubble removed. Image by EAG.

The next consideration for C-SAM is that the samples need to go into liquid, most often DI water. Typical DI water is 15 M Ohms out of the bottle and after a couple of days in the tank is about 2–3 M Ohms. The water should be changed every week at the longest or things start to grow. To impede the growth, I have used IPA and Listerine in 20 mL doses once a week to help delay the algae bloom.

The samples have to be able to survive the DI water, so no sugar cookies samples. Bubbles on the samples are an issue as they appear as round air gaps. For parts that do not have leads going into the packages, an IPA dip can be used to reduce the surface tension and prevent bubble formation. A simple squeeze bottle of DI water can be used to spray the submerged samples to dislodge bubbles as was done in the Figure above to dislodge the air trapped between the device and the PCB below. So, a couple of scans may be needed to ensure all bubbles are removed before saving the final data.

I have been asked which is better C-SAM or X-ray and the answer, which is a common answer in FA, is that it depends. We will go over X-ray in FA in the next chapter, but the short answer is that C-SAM is great for flat samples, that have a very thin air gap ($Z > 0.01$ microns) and is greater than 15 microns in diameter (X and Y) while X-ray is great for samples that have a significant change in density (1 micron solder particles where they should not be in a polymer for example). The two techniques are complementary and both have instances where they are the clearly superior choice in detecting certain types of defects.

The range in samples for commercially available C-SAM systems are small LEDs or ICs of 1 mm^2 up to 300 mm wafers and even 25-inch diameter shower heads that are four inches tall in the larger systems. Beyond these, specialty systems can be made for larger samples, odd shapes, and I have even heard of systems that rotate the sample in sync with the scans, to examine car tires for instance.

Chapter 8
X-ray imaging (2D and 3D)

X-ray imaging has several variants. Commonly in an FA lab, there is a 2D capability to quickly and nondestructively image a device to check for stray solder, broken wires, and obvious flaws in the interior of a device. The additional capability of 3D or CT (computed tomography) is growing more common. This enables devices to be imaged at many angles, usually every half a degree in rotation through 360 degrees, and then these images go through a software breakdown which enables the reconstruction of slices of the sample to be created. This typically has lower resolution than the 2D images by a factor of 7–30 but provides the slices without the interference of the rest of the sample. Something on the order of 700 slices are created in each orthogonal direction: front to back, top to bottom, and right to left. Each of these slices may be examined individually which removes the issue of having to look at the entire sample and getting interference from material above and below the area of interest.

Initially, when 3D X-ray became available as an analytical technique, I was very skeptical. In the vast majority of work, I would use the better resolution of 2D and avoid the long analysis times and added expense of the 3D analysis. Then, I had an issue with woven wire mesh and the 3D provided the answer when 2D could not hope to show the defect. The majority of issues are still more easily and quickly done by 2D but the 3D capabilities have proven that there is a place where it is the best option to provide the evidence of failure.

The X-ray image is essential a density map. The more the material absorbs the X-rays, the darker that area is in the image. So, seeing high-density material inside a lower density device is easy. For example, imaging solder inside a semiconductor package. Imaging aluminum wires, which is much less dense, is very difficult to impossible. For some reason, this convention is commonly reversed in 3D images with the lower density materials being darker. So, it is important to know the origin of any image.

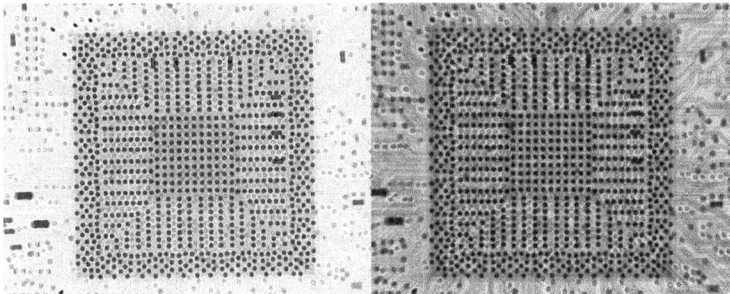

Figure 8.1: Standard and high definition of 2D X-ray images of a FC on PCB. Image by EAG.

https://doi.org/10.1515/9781501524790-008

Most X-ray tools in FA labs have a range of power they can use and the voltage is normally 20 to 160 kV. These tools are easy to operate and the images fairly easy to interpret. Obtaining a single image often takes less than 5 min. Adding on a 3D capability to a modern 2D instrument is often possible and can provide images on samples several inches in length and width. The best resolution is only possible when a smaller area is analyzed, approximately 5 mm³. When examining solder in a device on a PCB or solder bumps in a flip chip, both very common requests, an area up to 50 mm³ can be examined with sufficient resolution to spot missing solder, nonwetting, and solder runs.

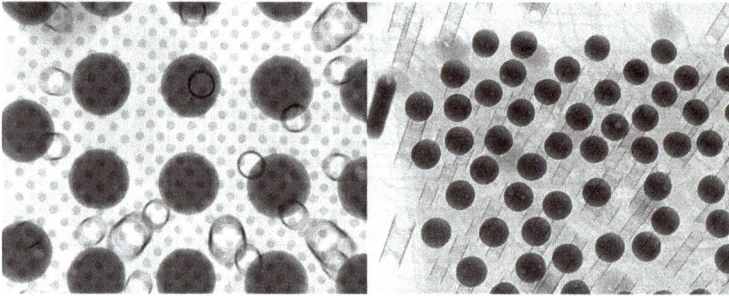

Figure 8.2: 2D X-ray images can be taken top down and at an angle. The images show solder bumps, balls, and vias. Note the voids, light spots, in the solder balls. Typical voiding in solder bumps/balls is 5–30%. Image by EAG.

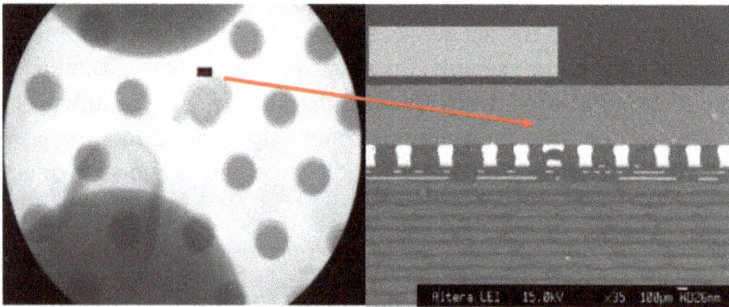

Figure 8.3: X-ray shows a solder issue which was then confirmed by cross section. Image by EAG.

The 3D capabilities are better on dedicated tools with sub-micron resolution possible, but a quality run takes somewhere between 4 and 18 h with some additional time for data analysis. The standard outputs are many layer files and these can be turned into 3D rendered images and even movies. Typical results will be ~700 images from right

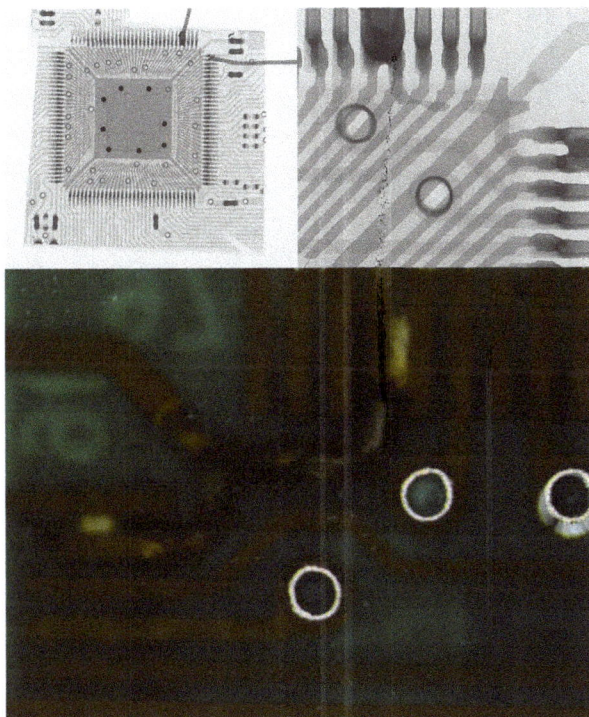

Figure 8.4: Examples of X-ray images showing substrate trace damage (nondestructively), which can then be viewed optically after some parallel lapping of the substrate. The optical image is dark field with IPA on the part. Note that the X-ray is top down and the optical is done from the bottom of the device. Image by EAG.

to left, front to back, and top to bottom. Going through all these images takes some time and is tedious, but can be done offline. Hopefully the defect of interest is fairly obvious, I can usually spend just a couple of seconds on each image, and someone may be working on some AI to help us out in the future.

The 3D data can also be rendered to provide 3D images and even movies.

Figure 8.5: Rendered 3D image of solder bumps. Image by EAG.

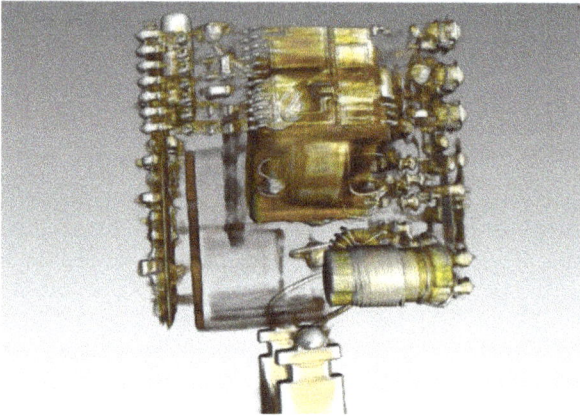

Figure 8.6: 3D X-ray of a GaN charger plug. Image by EAG.

Chapter 9
Time Domain Reflectometry (TDR)

I call the technique TDR the anti-paralysis tool. Often in FA, there is only a single device with a specific failure mechanism. This golden unit is alone in its ability to shed light on how this failure occurred and what steps should be taken to mitigate its occurrence in the future. When this failure is tied to a big project and the customer has eyes on the results, things can become tense. I have been in meetings with a failed packaged device and had the following question asked. "Is this a substrate, die, or interconnect issue? So, whose fault is it: fab, assembly, or substrate manufacturer?" Instantly fingers fly and everyone is blaming someone else. If the project will proceed or be cancelled depending on the result, the strong inclination of the project owner is to freeze and not proceed if the results cannot be guaranteed.

As the FA person, you have to not care what the answer is. You cannot have a vested interest in the outcome. Even if you are not swayed by such things, there will be a perception that you are if you do not hold yourself to a stated position of neutrality. Also, the pressure from the interested parties can be quite intense.

So, if the failure is not in a known location, there arises the issue of what can we do? If C-SAM, X-ray, and IR do not indicate where the issue is located then how to proceed? There are slight changes depending on if the device is a wire bond or flip chip (FC) part, but the issue is essentially unchanged. If we expose the die for EMMI/OBIRCH/IR then we may affect the strain on the device or remove contamination on the die surface and the part may recover without direct evidence of the cause. If we instead remove the substrate by polishing, and the die then tests as good, then we again have no direct evidence of the issue in the substrate.

TDR addresses this problem. A good part, the failing unit, and a known good substrate are tested. Often the known good substrate is made by polishing a good unit to either break the wire bonds or remove the die for a FC part. These units are tested by placing each into a probe setup that contacts the solder ball of interest and an adjacent ground plane ball. A pulse of electrons is put into the part and the reflected pulses measured. At each interface, a partial reflection occurs. An example of TDR curves are shown in Figure 9.1.

In this comparison of the results from the good device, substrate, failing device, and open socket, it is possible to determine where the failing part deviates and therefore where the defect is located. Between the open socket and the end of the good substrate, the issue is in the substrate, between the end of the substrate and the end of the scan it is in the die, and if the divergence is right at the interface of the two it is in the attach (bond wires or solder bumps). This information allows the FA to proceed as knowing the location of the defect enables the correct choice to be made on removal of the die or substrate to continue the search for the defect with confidence. An open failure observed by TDR that is confirmed in cross section is shown in Figure 9.2

https://doi.org/10.1515/9781501524790-009

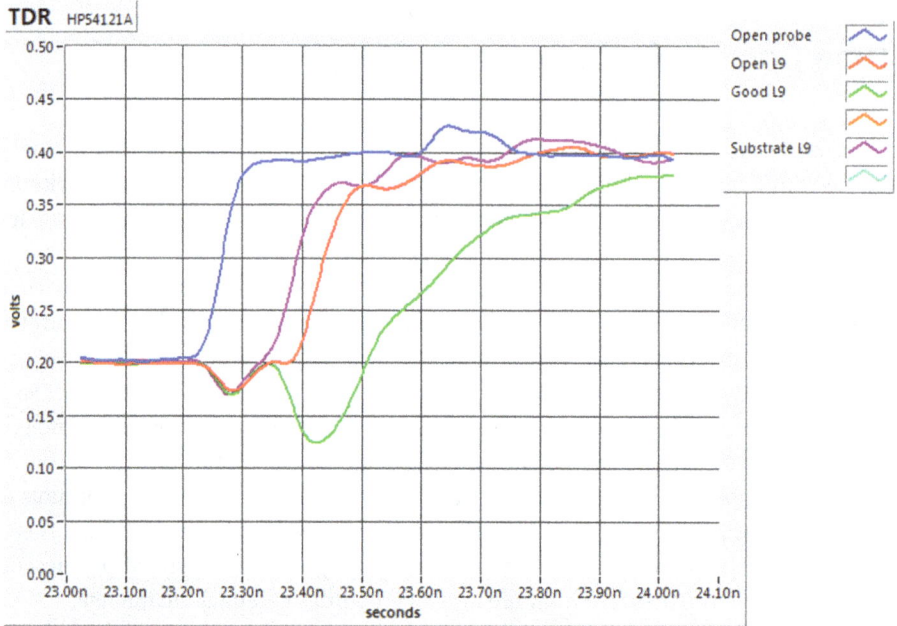

Figure 9.1: TDR waveforms of reference (open) blue, known good substrate purple, and failing unit red, and known good unit green. In this case, the FU deviates after the KGS so the defect is in the die. Image by EAG.

Figure 9.2: Cross section of an open failure that was initially diagnosed with TDR. This is a cracked via chain in the substrate. Image by EAG.

While TDR tools are not terribly complex or expensive for a typical tool, they can provide crucial information on how best to proceed with an FA. The need for a ground and the pin to be tested can cause issues if the ground is far away. Some systems have separate probes, but the tools I have used all have the ground and test probe locked together. In these cases, the ground may have to be moved by using metallic tape or other ways so the proper contacts can be made.

The curve tracer, the brains and muscle for TDR, will determine what timing, spatial, resolution can be achieved. The faster your box the better the resolution that can be achieved. Some devices are so small that the systems I have most recently used cannot really separate much and on some it is useless as the entire path is too short.

For FA labs that will analyze the same parts over and over, it may be worthwhile to create a set of standards. By this, I mean taking known good parts and breaking a known trouble path in different locations, placing a break in the substrate traces after each via for example. This will enable rapid location of any defects without the need to physically de-process the failing substrate.

Chapter 10
Selecting the path before destructive analysis begins

After the nondestructive techniques have been exhausted a path should be planned out for the remainder of the FA. This does not mean all remaining steps are locked in, but that a path of if-then statements is agreed to ahead of the work. This avoids the "Let's just try this and see," error and gets all invested parties aligned to avoid blaming and bitterness later. When such a plan is made it also allows the work to then proceed quickly instead requiring a meeting after each step to then decide what to do next. I have found that the process goes best with a plan and regular updates by email. This has the added bonus of allowing the report to be written as the work goes along and at the end it is an easy finish instead of trying to do the report all at the end.

Typically, the earlier steps (optical microscopy, C-SAM, X-ray, electrical tests and IR) have either identified the location of a defect or not. In the case of a localized defect, the choices are now down to the method(s) to be used to expose, image, and analyze the defect.

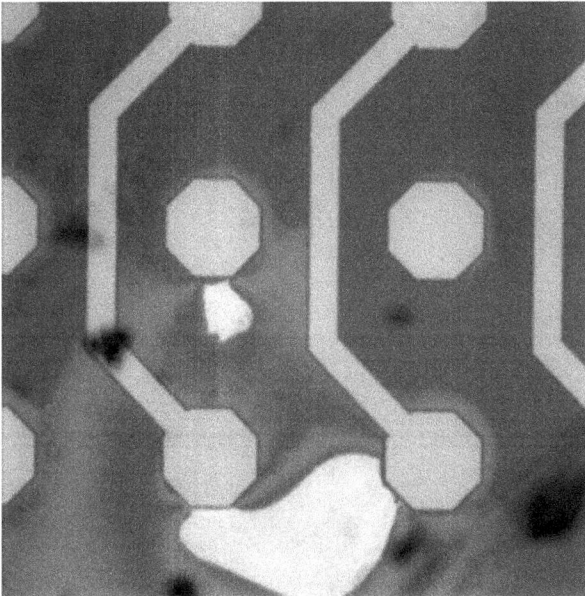

Figure 10.1: IR image showing melted metal causing shorts in an IC. P-lap or cross section next? Image by EAG.

https://doi.org/10.1515/9781501524790-010

In the case where no localization of the defect has occurred, there are several choices that must be made. First and most often the best choice is to end the FA with no defect found. This is often the best choice as it does not waste the time of the FA lab and the others involved in the process. Second, stress the part to make the defect worse and hopefully easier to find without inducing new and unrelated defects. This can involve running the part at Imax or a little higher current/voltage at elevated temperature or cycling the part on/off. When the leakage or resistance or whatever the failure indicator is reaches a specified level, the FA can resume. Sometimes, the level is not reached and the set time or number of cycles in used instead. Never agree to do an FA to "look around for defects" as there are normally many anomalies on any given device. Almost all of them have no effect on the device performance. A study of such anomalies is fine, but do not do an FA this way or you will spend endless hours searching and explaining things that are unimportant, just don't do it.

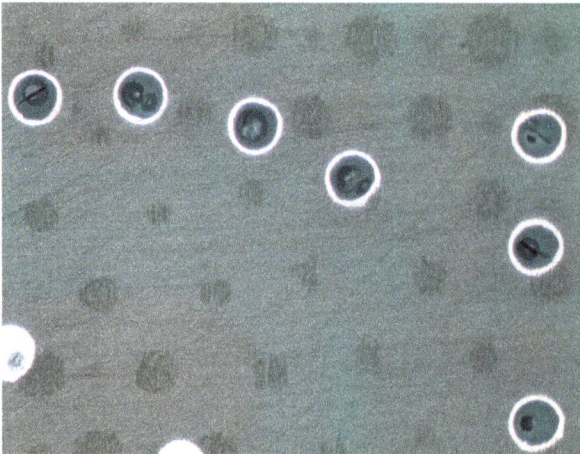

Figure 10.2: Defects in filled vias. This does nothing to the device performance. Image by EAG.

Now if a defect can be localized the decision on next steps should be laid out and agreed to by the interested parties. This makes the future steps easier and gets all the parties on the same page. This is also the time that all parties can give their knowledge and experience to the effort. It is best if there is a report with images and history of the steps already taken provided at this time.

The path after localization can vary quite a bit. Sometimes, the product engineer will understand the issue from just this information and the FA will be closed. This is rare as there is always someone that wants to see the defect directly.

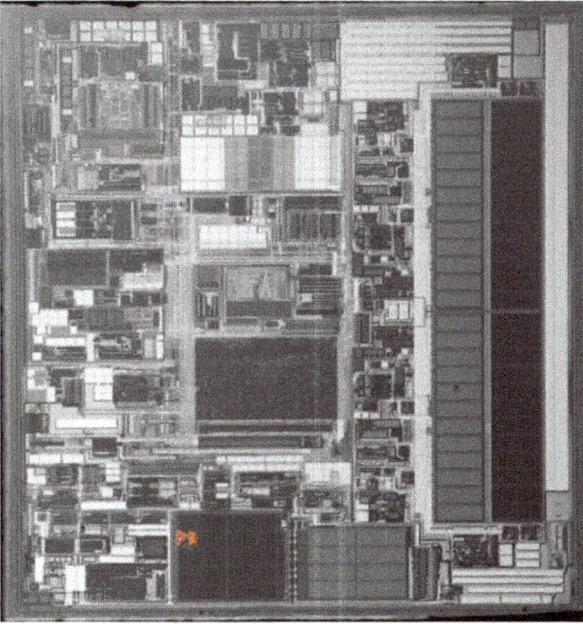

Figure 10.3: Localized defect area observed by XIVA. Image by EAG.

The most common paths involve removing material from the sample and looking at the defect location. The next couple of chapters will go through these processes.

Chapter 11
Destructive work

Once the non-destructive testing (NDT) is completed, the fun of doing the destructive work begins. This is sometimes referred to as physical failure analysis (PFA). This is a point in the FA process when a bit of careful thought, planning, and getting the buy in of the concerned parties will save you from difficulties later. Getting the go ahead for doing a decap or confirming the exact location of the cross section and receiving a reply email confirming the go ahead ensures the step is agreed to and correct as far as all parties are concerned. Without this confirmation all mistakes are yours alone. I strongly advise this step because a quick check will save you hours of work and the destruction of valuable samples that may provide solutions to your company's issues. The request for idiot-proof instructions should not bother any of your clients. I have sent images of samples asking for red lines to be put into the document telling which areas are to be sectioned and arrows for the direction of polish, because "row four" is not enough to be sure it will be done correctly.

This is a good point to mention setting rules for your analysis procedure for the lab. For instance, all images of devices should be done with pin one in the lower left (LL) corner. Cross section images should always be done with the top of the device at the top in the images. Of course, you can take a different image, but it must be labeled in the image. This cuts down on decisions during the FA and makes images easily understood later. So, on a call with a customer a month later, I know the orientation of the image because all lab images are done a certain way. Each image file should also be named in a way so that the sample ID is included.

You do not want to explain in the next customer meeting how you did not check to get confirmation and then cross sectioned the one-of-a-kind sample on the wrong row of solder bumps and the area of interest is now gone. I have had instructions from engineers that I have double checked and were found to be in error. I usually send to the requestor the top-down X-ray of a unit with pin one in the LL and the section line in red with an arrow pointing the direction of approach. This way there is no possible misunderstanding. Stating in the email request that this is top-down X-ray instead of using a layout drawing makes it impossible to misunderstand. Sometimes groups feel it is better to blame FA for destroying a sample than to deal with the failure found. Do not let them use you in this way. Get them on the hook with an email stating the go ahead before you do any destructive work.

It is best to take some optical images of the sample(s) before starting the destructive process. It is also advisable to mark the sample with permanent marker,

https://doi.org/10.1515/9781501524790-011

diamond scribe, or laser marker if the orientation of the sample may be lost as the destructive work proceeds. This is usually needed only for delayering and sometimes cross sectioning.

This is also the time to identify the samples to be further analyzed, obtain any desired practice samples, and understand what the expected possible results are and what the results will mean. This way you can better understand what you are seeing during the process and know what is of importance. Stopping and taking images of every anomaly is a waste of time if you know what the real concern is for a device and those other images can be skipped. This will also enable you to write a report highlighting the issue that is of concern and detailing anything found that matches the concern.

In destructive work, you may need to involve the customer in parts of the process; such as when you find something unexpected before you reach the designated area of interest. Be sure to get an image and send an email to the customer so they can better understand the FA. In this way, you help the effort to find the root cause problem and hopefully solve it. Nothing is more frustrating for the FA team than a well-done job that shows clear results and then operations does nothing and the same failure mode shows up in the next week's inbox. Involving operations in the FA and the results help to mitigate this issue.

If the sample needs to be taken apart, for example removing a device from a PCB, be sure to let the customer know before doing the operation. If electrical leakage is the issue, they may not want any heat applied to the sample, so this will influence your course of action. If a device needs to be removed from a PCB, we usually use heat if this is not a concern, it is fast and easy. But if heat may alter the failure, then we can cut out the portion of the PCB containing the device of interest and then polish the PCB away. Be sure to cut far enough away from the device to keep it safe. Diamond saws are the best for this work as they introduce the least amount of heat, but they can only handle smaller samples typically and may introduce additional complications from contamination with the cutting fluid. Band saws are very violent and may cause concern for your customers as well as introduce extensive heat. The Dremel tool is a must in all FA labs and usually can do what is needed in almost all occasions. Do whatever is safest for the sample and easiest for you. Be sure to check that cutting up the PCB is acceptable before doing surgery. During any mechanical cutting operations keep in mind the heat transfer from the cutting region into the region of interest from any metallic connections.

The Dremel tool is nice as any angle can be cut and the sample dimensions can be quite large.

Figure 11.1: Dremel tool used to cut the top line to release sample from PCB. Subsequent cuts on the now smaller PCB section are completed with a diamond saw. Image by Dan Sullivan.

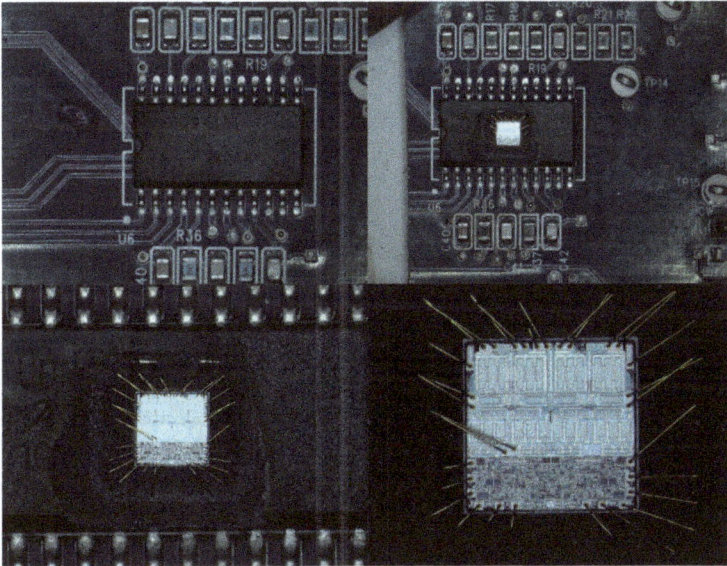

Figure 11.2: Decap is often required to expose the die for optical inspection and fault localization in the die. Image by EAG.

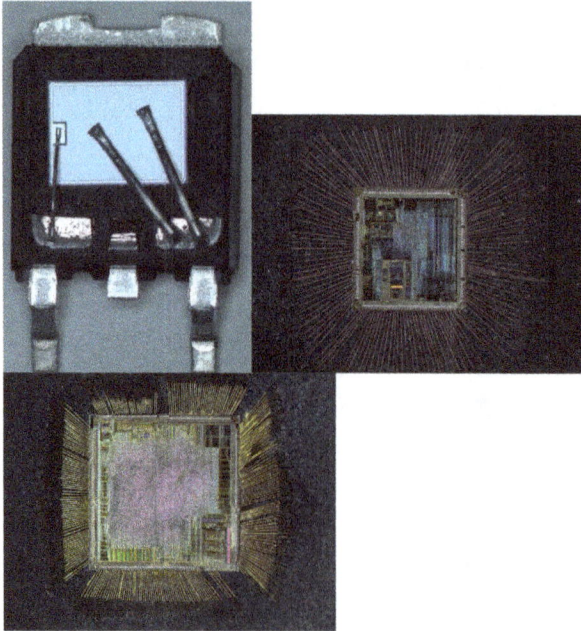

Figure 11.3: There are many different types of packages that each requires different recipes for decap. Examples: MOSFET with Al and Au bond wires and very thick encapsulant. Devices with copper wire (right) and gold wire (bottom). Image by EAG.

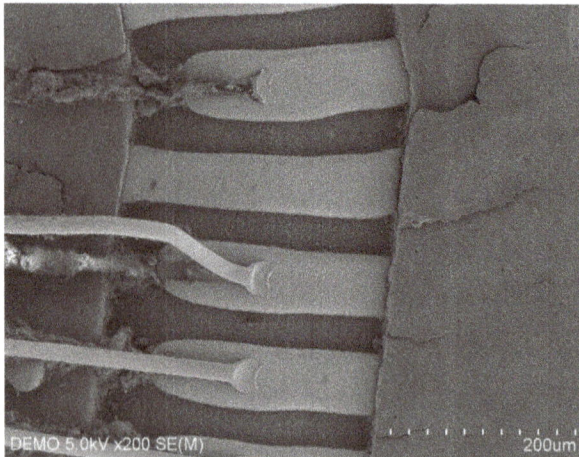

Figure 11.4: Laser ablation can also be used for exposing the second bonds to the substrate. Here, two stitch bonds are exposed; note that the top one is raised up and near breaking. Image by EAG.

Figure 11.3 shows a device as received mounted on a PCB. The device is dec-apped and the zoomed in images show the die and bond wires. FA often requires access to the interior of a device. This is done by decapping to expose the die and allow inspection and electrical probing. Decap is most often done with nitric and sulfuric acids at elevated temperatures. This is part science and part art. The process should keep the device electrically intact if possible. Laser ablation helps here by precavitating the decap area which leads to higher success rates.

Chapter 12
Electrical Failure Analysis (EFA)

Electrical analysis is done only on electrical devices or systems, duh. There are electrical tests for other materials such as breakdown voltage or resistance, but those are fairly straight forward and the results are very easy to understand and therefore are not covered here.

For ICs (dies and wafers), PCBs, and packaged devices, a contact, or multiple contacts, can be made. This enables application of voltage and input current or monitoring of the outputs. This can be done either by attaching wires (usually by soldering) or probes can be landed by hand to the proper points or a microprobe station can be used to put very fine needles in contact with probe pads, wires, and probe points made by the FIB. The probe needles come in a wide variety of sizes and flexibility. I find that 7 um ones work best for general use. These are placed over the location of interest and lowered under a microscope until the probe touches down. Examples of this are shown in Figure 12.1. Getting this right is a skill that requires good eyesight and lots of practice. It is easy to go too fast and force the probe onto the part which will cause it to slide on the surface, causing a scratch. In the wrong place, this scratch damages the die and your FA may be done with a result of destroyed in FA, not the optimum outcome.

Typically, there are two probes landed to enable a voltage to be applied on one that runs through the die or packaged device to the other probe where the output voltage and current are measured. Often, this is to check for a short or leakage between two pads that are not supposed to be electrically attached. So, in a good part, the current will be zero up to some breakdown voltage and if you get to the breakdown voltage, it will be a very sharp rise in current past that. Here, it is important to have the specification of the device from the customer. This includes Imax and Vmax. The current should be clamped so the current does not go too high possibly damaging the device. The purpose of an FA is to find the root cause of the original failure and not just random damage caused during FA.

I have worked with an engineer who when they could not find a hot spot on a device in the specified voltage range would just raise the voltage until one appeared. This behavior is very bad and should result in being fired on the spot when discovered.

There are parts that need additional probes to ground some pads or hold them at a particular voltage and for more complex issues it will sometimes be necessary to run the device through a startup process to get into the proper mode to observe the failure.

In all of these cases, the IV, current versus voltage, curves may be collected. In this way, the characteristics can be compared to a good part and/or the theoretical expected results. Typical issues are: shorts, leakage, opens, and intermittent variants of these. Examples of these collected curves are shown in Figure 12.2.

https://doi.org/10.1515/9781501524790-012

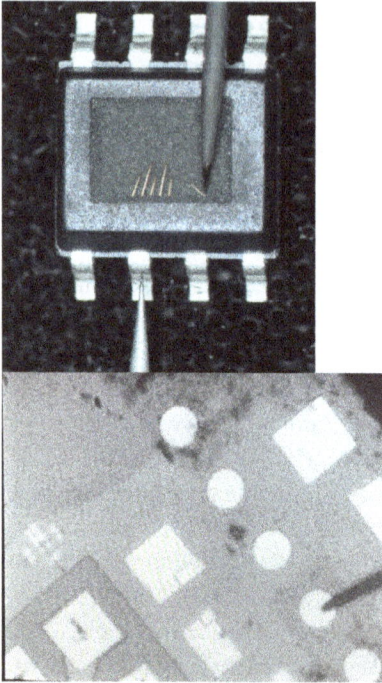

Figure 12.1: Probes touching bond wires, leads, and on bond pads directly on die. Image by EAG.

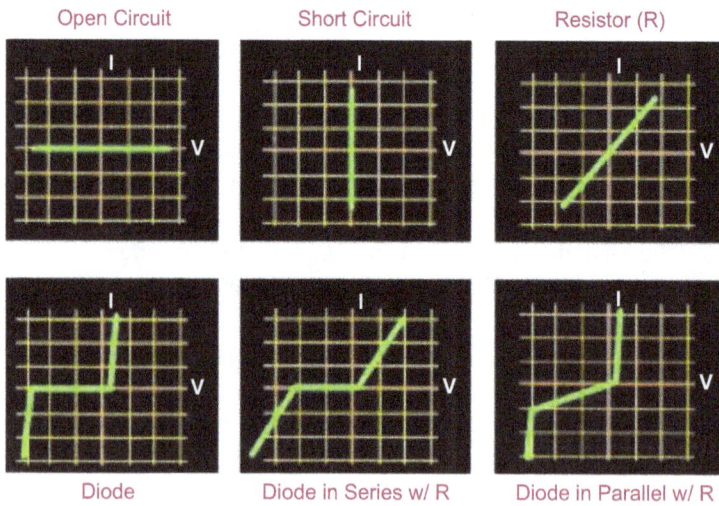

Figure 12.2: IV curves showing shorts, leakage, and open. Image by EAG.

The failures are checked to ensure that the part has not recovered, the sample is the correct part (accidental swaps have been known to happen in the lab, from the customer, and even in the QA inspections), and to ensure the instructions are clearly understood and the results match the initial issue. For leakage and shorts, it is often a good idea to bake the sample at 100 °C for at least 12 h to see if the part will recover. If the part does recover, the likeliest causes of the issue are moisture ingression, dendrite growth, or stress in the part. Also, if the part recovers partially, this gives clues on the most likely issues.

In this example, a wafer has several die that are failing test for leakage. The customer usually provides a wafer map and the test results. In olden days the failing die would be marked with an ink spot, this is uncommon now. Typically, the worst failing die or a couple of bad ones are selected and the wafer is placed onto a probe station and probes landed on two or more pads. The typical issue is one pad is leaking to another that should be open line (OL) to it. The IV curve of this connection is then run from 0 to the specified voltage of the part. This can be zero current all the way if the two pads are not supposed to be connected at all or if they have a turn on voltage the current should rapidly increase at this voltage. The current is usually clamped so a specified current is not exceeded to avoid damaging the part during the test. This is typically 1 mA. In this way, any set of pads can be tested and IV curves for any device can be obtained. Extra probes can be landed if a voltage on another pad needs a certain voltage or a series of operations need to happen in the device to get it into the correct state to observe the failure. This may require the use of a device under test (DUT) board if the setup is complicated or requires more than eight probes.

If no leakage is found on the part, there are several possibilities: the part is set up incorrectly or the wrong pads have been probed, the wrong part is being tested, or the part has recovered. A check with the instructions or the customer to confirm the set up and the identity of the part will usually solve the first two issues.

The possibility of parts recovering varies depending on the failure mode and the conditions that caused the failure. In stress tests that involve damp heat, it is advisable for leakage and shorts failing devices be baked out at 100 °C for 4 to 24 h and then be retested. Moisture ingression alone can cause these failures and the bake will remove the moisture and the part will recover. Other causes of failure that recover from baking are the formation of dendrites (typically with Ag die attach) and Sn whiskers. An example of dendritic growth observed with 2D X-ray is shown in Figure 12.3. Both can be broken up by a bake.

Performing the bake before proceeding will save you time and effort trying to find water in a device. The best way to proceed after a bake recovery is to recreate the conditions of failure again and look for the failing signature. If this is a water issue, it will repeat the same way. If it is dendrite or tin whisker growth, it may or may not. The part can then be decapped and can be searched for signs of dendrite and whisker growth. It is best if the part can be decapped without wet chemistry as the silver dendrites tend to

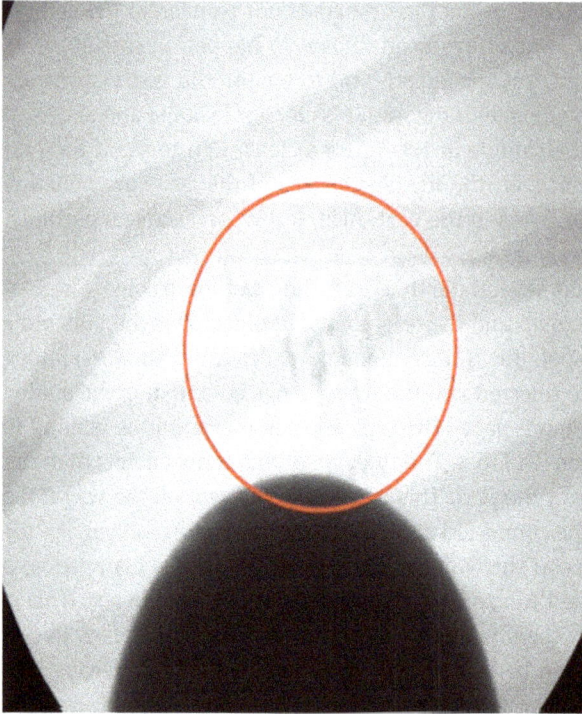

Figure 12.3: Dendritic growth observed by 2D X-ray after HAST. This is Cu dendrite growth. Image by EAG.

get etched away and the tin whiskers get broken. I have been able to find both by using a plasma decap and once by very carefully lapping away the encapsulant.

A final electrical signature that can be very frustrating is the intermittent failure. I have seen this many times now so it's not such a difficult issue anymore, but the first few times this type of failure is encountered, it can be very difficult to deal with. These are usually an open or high resistance that recovers when pressure is applied to the device or the part only fails at high or low temperature. This is due to a contact being made intermittently when the pressure is applied to the device or materials expanding or contracting with temperature to make or break contact. This is typically seen for solder or ball bond cracking or nonbonding. An X-ray image can sometimes reveal the issue, but small cracks in the solder or ball bond lift can be very difficult to detect in the X-ray images.

A decap and direct inspection of the ball bonds will often reveal the lift as the encapsulant that was holding the ball in place is removed and the wire easily lifts. For the case of the solder, the non-wets are often seen by X-ray, head on pillow, and the cracking due to stress or large growth of the intermetallic can be seen in a cross section.

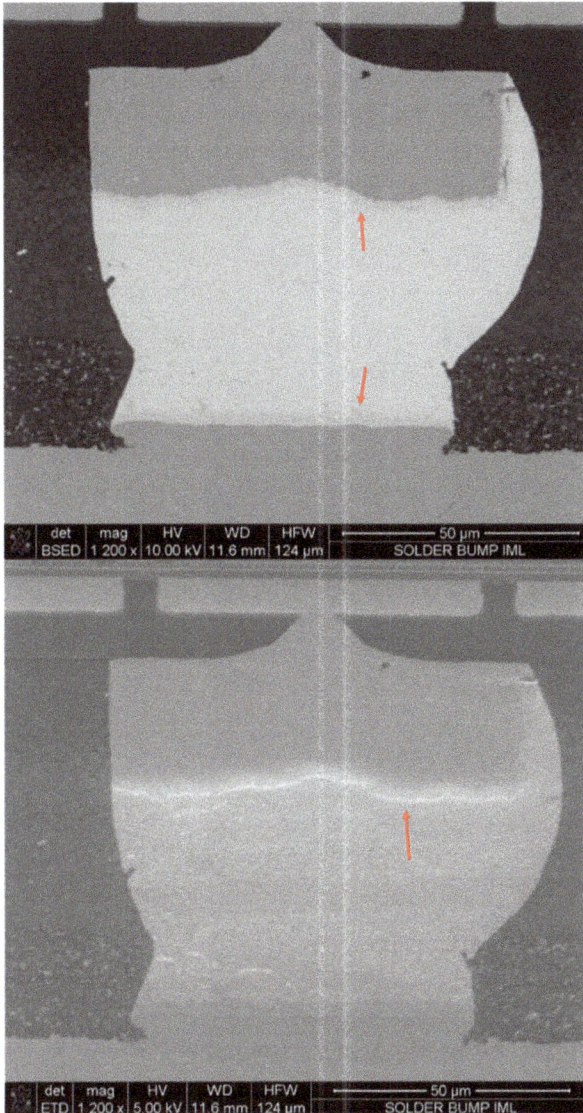

Figure 12.4: Image of intermetallic growth in the solder bump/bond pad interface. The backscatter image and the normal SEM image are of the same location. Note that EBSD makes the IML much easier to see. Image by EAG.

In cases of the incorrect phases being made, we see purple or red plague in which the improper phases are brittle and crumble. In the case of excessive IML growth, the materials get used up and create Kirkendall voiding or stress cracking.

The SEM in normal mode may show the IML. The backscatter mode increases the contrast based on the Z value of the material makes the IML much more distinct. Images of IML growth in both modes are shown in Figure 12.4.

Chapter 13
EMMI/OBIRCH/IR

The group of analytical techniques that are used to localize electrical defects in semiconductor chips make a long alphabet soup: LIVA, TIVA, KIVA, etc. But the three that I have used most often are light emission microscopy (LIMS) which I call emission microscopy (EMMI), optical beam induced resistance change (OBIRCH) which is the family with TIVA, KIVA, and LIVA, and infrared emission microscopy (IR) which detects heat. All of these require direct access to the die, so a decap or delid process is often required to gain this direct access. The power inputs must be readily available for either direct probe to bond pads, solder balls, or other connections to the die. The die being analyzed must also have a direct line of sight to the detector and for OBIRCH the probing laser. This can be to the front surface or back surface of the die. The backside analysis is most often done for flip chip devices or for die that have large metal areas that block the underlying circuitry.

In the case of the back-surface analysis, the die must have a polished finish and no metal coating on the backside. Sometimes the die is too thick, is too highly doped or is rough or has a metal coating. In all of these cases the die can be thinned and polished to enable the technique to work at all or improve the imaging and signal detected. The process to do this thinning depends on the clearance around the die. The preferred method is for a fully cleared die and is a universal polish. This leaves a mirror finish and can routinely end with a die thickness of ~50–100 microns. The alternative method uses a milling process that is good but leaves slight waves in the die surface that affects the image quality. The mill method does have the advantage of allowing smaller areas to be thinned and polished leaving the surrounding area untouched. In this way, the crucial parts of the substrate can be left intact and the stress on the part is altered to a minimal degree. This is also done on wafers where only a single die or section of a die is of interest.

There are several manufacturers of EMMI/OBIRCH/IR instruments, but I only know of one that makes all three on one tool. This tool is great in that the setup is the same for a sample and in the analysis nothing changes but the detector and the software being used. This tool is expensive but having all the capabilities in one tool instead of two or three is very convenient and likely actually reduces the cost compared to having two or three separate tools.

Each of these techniques works easily for a high power, light emitting leakage. Most issues are not this easy. The standard steps are: EMMI then OBIRCH and at last, IR and then only if there is greater than 30 mW of power. In all cases, an IR image of the device is obtained and the signal detected in the analysis is then overlaid on this image.

For EMMI, set up the probes and check for light emission starting at 0 V and slowly increasing up to the maximum voltage allowed. Once emission spots are

https://doi.org/10.1515/9781501524790-013

observed, if they are, then the process is stopped and the images are obtained. A typical image showing EMMI spots is shown in Figure 13.1. It is important to do this from low magnification up to the highest magnification so it is easier for the physical deprocessing steps later. If the maximum allowed voltage is reached, it is important not to exceed this as finding the damage you just induced on the chip by blowing things up with over stress is not very helpful and is in fact very unhelpful if it is not recognized as induced in the FA and the results from it will be misleading. Wasting the fabs time hunting false failures is not a good use of resources.

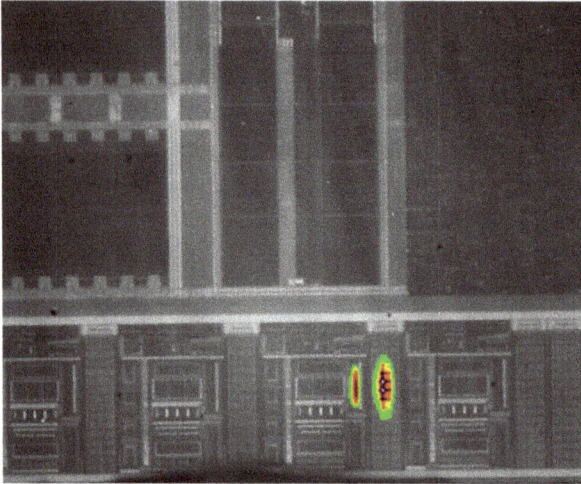

Figure 13.1: EMMI spots observed on a die. Image by EAG.

OBIRCH uses the same connections to the device as EMMI. The image of the circuitry is obtained in the same way. It is the probing and detection that are different. In OBIRCH, a laser is scanned over the area of interest while the input of interest is powered. The current or voltage is held constant while the other is monitored (this is where all the names come from). When the localized heating by the laser causes a change in the resistance on the device, the monitored property will change. An example of this is shown in Figure 13.2. This usually does not happen in good parts. It does happen often enough that a good part to compare with is recommended. The location of the resistance change is noted on the image with a red or green spot depending on the change being negative (red) or positive (green). The resolution is quite good, being a couple of microns.

These techniques are often misnamed as hot spot analysis. These techniques replace the use of liquid crystal which is where the name hot spot comes from as the liquid placed onto the die crystalizes in a temperature range just above room temperature. So, the hot spot turns black as the voltage/current rises which heats the material. The common trick for hot spot analysis is to raise and lower the power (usually by V)

Figure 13.2: OBIRCH spot. The location of the laser when the resistance change is detected is overlain on the IR image of the die. Image by EAG.

to get the temperature at the failing spot to go just above and then just below the crystallization point, the spot will pulse and can be easily found, most of the time. These liquid crystal materials are carcinogenic and sticky and a pain to deal with. So, the replacement with the EMMI/OBIRCH/IR tools is very welcome. If you are working with liquid crystal, it is very important to deal with it as a hazardous material.

These techniques, unlike liquid crystal, can also be performed from the front or backside of a die which is very helpful for flip chips and devices with many metal layers and/or areas of high metal density. If the die is very thick, highly doped, or has a rough finish, the die will need to be thinned to allow these techniques to work. Typical silicon die are ~700 microns thick. This can be thinned down to ~100 microns easily and with some expertise and can go down to 50 microns without too much risk. If there are no obstructions around the die, this can be done universally on an autopolishing tool with very good results. If there are issues with obstructions around the die then a milling tool can be used. These usually provide an adequate polished surface but not as good as the universal polish.

The IR camera used for the detection of IR heat in the device can also be used simply to image the circuitry, an example of each are shown in Figures 13.3 and Figure 13.4. This is very helpful for inspection of flip chip die for damage from dicing operations that can cause cracking at the edge of the die. The resolution of the IR is about 3 microns but large cracks that penetrate the seal ring or reach the active parts of the die are often much larger than this.

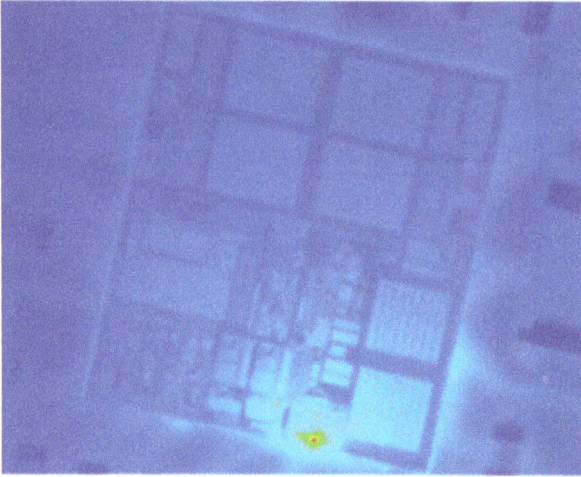

Figure 13.3: IR hot spot raw data image of the die. Image by EAG.

Figure 13.4: IR imaging through the backside of a Si die. Image by EAG.

At one company, the FA group used IR to find scalloping, round cracks coming in from the edge of the die that hit and often penetrated the seal ring. This issue was observed for several weeks on dozens of parts submitted to the lab. In cooperation with the other engineering groups, we were able to show that the location of these cracks matched up with certain objects in the streets, the area between the die on a wafer. In the streets, the designers had placed a variety of objects used to check dimensions and electrical characteristics of the wafer during and post process. Some of these objects were metal. When the dicing saw struck particularly hard objects, it caused cracking in the area and sometimes was severe enough to penetrate into the die and cause these failures. Images of the streets and metal objects in them are shown in Figure 13.5.

Because of this FA, the layout was changed and these metal markers were removed and the issue went away. This improved the yields a couple of percent which improved the devices bottom line far more than the costs of the FA.

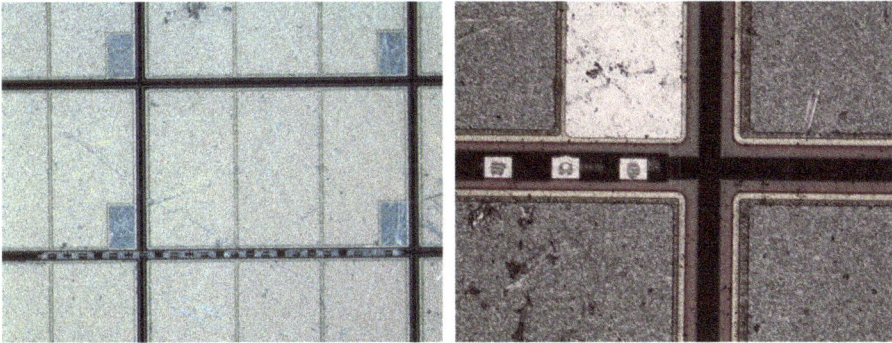

Figure 13.5: Image of the streets (area between die on a wafer). Note the objects in the streets. Image by Dan Sullivan.

Chapter 14
Cross sections (X-sections)

Cross sections are just what they sound like: the device is cut from top to bottom to allow inspection of the materials/layers inside. How sections are done depends on what the sample is and what materials are involved [1]. The four main types of x-section are: cleave, mechanical polish, ion mill, and dual beam focused ion beam (DB FIB). All involve some type of inspection during and after the section is complete. This is by means of optical microscopy (OM), scanning electron microscopy (SEM), and transmission electron microscopy (TEM). Each sectioning and inspection technique has advantages and disadvantages. In some cases, multiple methods may be combined in an FA [2].

Cleave work, sometimes referred to as XSEM, is used on semiconductor wafers and die. The sample is cleaved to look at general areas to inspect interfaces or measure layer thicknesses.

To hit specific targets, a microcleaver can be used for greater accuracy in the location of the cleave. I know of one such tool made by Sila that has an accuracy around 1 micron. This approach is most often used to measure critical dimensions for a fab monitoring or process development work where speed is the primary need. The downsides are that the sample has to be prepared to a small size before the microcleave and the location must be visible in some way to align the cleave. A hidden location can be marked by laser marker or FIB cut prior to the cleaving. Cleaving is not difficult but some experience is preferred so the cleave is nice and not jagged or worse shattered.

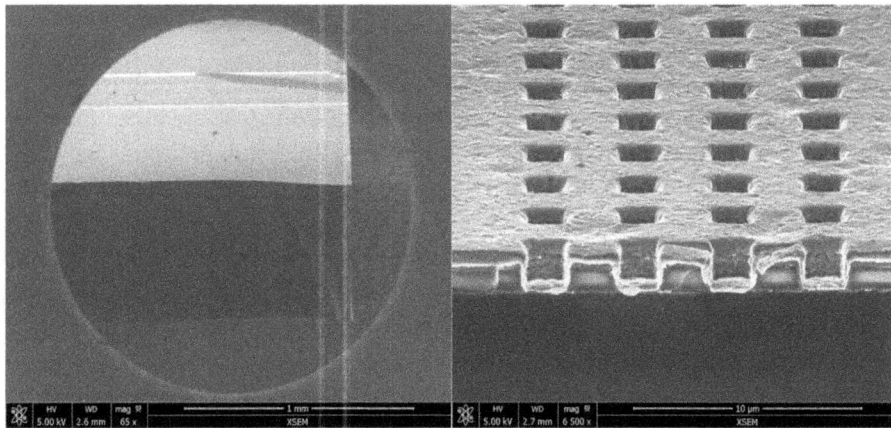

Figure 14.1: Titled images of a cleaved cross section. Cleaved faces are the forefront surfaces. Image by EAG.

https://doi.org/10.1515/9781501524790-014

Mechanical sections come in two general types: die only (silicon or other substrates – GaAs, Ge, InP, etc.) and everything else. During the die only sections, the sample is typically encapsulated with a two-part epoxy that is used to secure the sample between a piece of silicon support and a glass cover slip. This is for support in handling the piece while allowing visual checks to be made on the progress of the section.

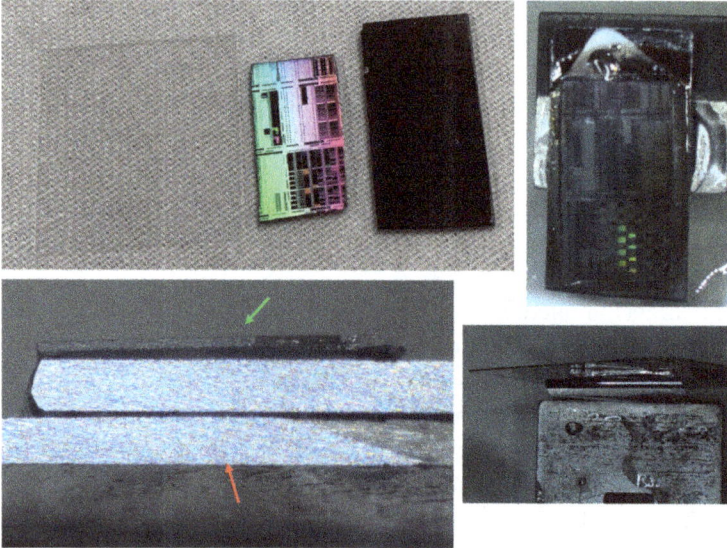

Figure 14.2: Images of a die only section: Pieces pre-assembly, top and side, and the final polish face with cover glass on top (green arrow), Si support on bottom (red arrow), and epoxy between layers. Image by Dan Sullivan.

The general process used for preparation of cross sections of die is illustrated in Figure 14.2. The cover glass, sample, and support silicon piece are glued together with epoxy and cured. This piece is then mounted on a stub and polished until ready for inspection in the SEM. This stub fits on a tripod polish holder and the SEM holder.

The die only section epoxy uses a hot plate or oven as an external heat source. If the heat is too high or left for too long then the epoxy can turn a very dark brown, making visual inspection through the transparent cover slide much more difficult.

For samples involving more than just the die, the potting encapsulant will depend on what you prefer and what your sample can take. Although a detailed treatise on encapsulation is outside the scope of this book, it is informative to know that there are a variety of materials and methods for encapsulating your sample for subsequent polishing steps. The typical encapsulation methods employed are compression (cold or hot) and cured liquid materials. The materials available include epoxy,

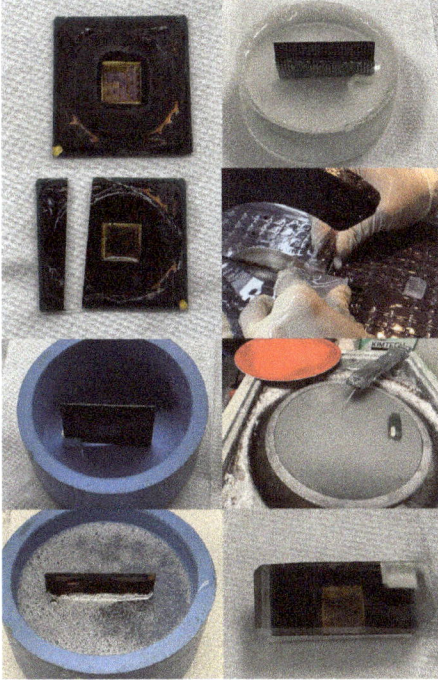

Figure 14.3: Cross section cutting, mounting, and polishing for packaged devices. Process starts with cutting sample to size, mounting, cutting mount to size, and polishing, resulting in final polished cross section mount for further imaging and analysis. Image by Dan Sullivan.

acrylic, phenolics, and polyesters, all of which have advantages and disadvantages. We have also encountered several cases where the ultimate cosmetics of the finished cross section must be good and the encapsulant must be very clear to produce a "dog and pony show" part to show investors and other non-technical people. The general process for preparation of samples that include more than just the die is shown Figure 14.3.

We have used a variety of encapsulants and done a DOE at one point to find the one system that stays clear, does not stink, and does not reach an unreasonable temperature. Unfortunately, several samples over 20+ years of doing this have recovered from short/leakage or have an indirect failure created when potted, so for most samples, we have abandoned all epoxies that have a curing exotherm of over 100 °C. The one we use for non-die only samples has a curing exotherm of 80 °C and takes ~8 h to set, but it does not melt solder or warp packages. We figure it is better to have the good results tomorrow with a slow cure than ruin the sample today.

Now, some may say that you should always get the best possible finish every time and in a perfect world without time constraints, and they are right. In real world situations, that approach is short-sighted resulting in needless waste of time and material resources. If the purpose is just to determine the number of metal layers in a device, then you may just cut the part with a saw and take a look; no need for fancy polishing. The purpose of FA is to answer questions and not to make pretty images for their own sake. Get the area of interest clean and free of artifacts and move on. In addition, you can return to the sample and polish to a better finish if it becomes pertinent to the final results while compiling the FA report. This is only true if FA does not require additional polishing further into the sample.

Once the general area of interest is mounted and roughly polished, fine adjustments of the polish depth can be achieved by optically observing through the polished "side window" and with X-ray imaging through the mount. This process is illustrated in Figure 14.4.

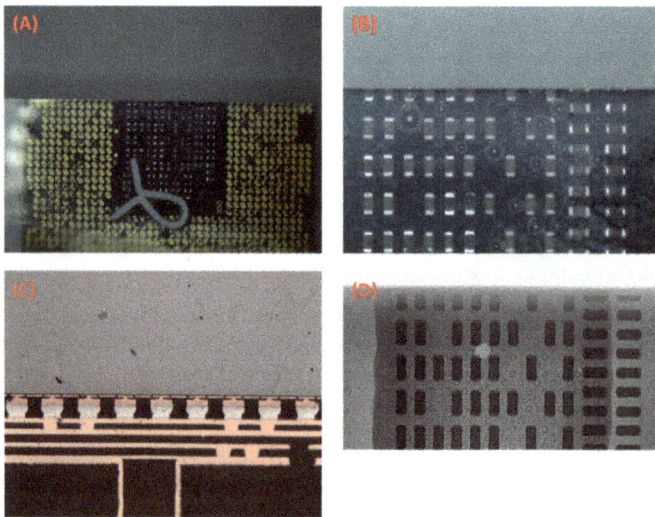

Figure 14.4: Top down low (A) and high (B) magnification optical images of the cross-section depth viewed through the side of the mount showing depth of polish into the mount, X-ray image through the mount (C), and the optical image of the mount cross section at that depth (D). Image by Dan Sullivan.

The die only sections are typically done with diamond-impregnated films. The rating system for these is 30 down to 0.1 (the number is the diameter of the diamond particles in microns) followed by various polishing cloths with diamond or silicon colloidal suspensions. Everyone has a secret process they use and it will change depending on the materials and the desired quality of the final surface. The films are color coded so using the wrong one is easily avoided. Colors of papers and solutions mentioned below assume the use of Buehler consumables; however, there are many alternatives and their color to particle size designation can vary. I also write the grit size on the backs of the films with a permanent marker. It is better to over prepare than to have to write DFA (destroyed in FA) in a report.

These films have a rough side and a smooth side. The rough side is slightly darker and less reflective so identification is fairly easy; by writing the grit size on the back, it's bulletproof. I had a senior person in one lab who managed to train a junior person on doing these cross sections. The senior person, it turns out, is not a very nice person and trained the junior person the wrong way around on the two sides of the films. I caught this when checking in on the lab. The junior person asked why the section was taking so long and when I took a look, the film was up-side down. The senior person claimed that the junior must have misunderstood and added that they were too stupid to do the work. The senior person never did any-more training in my lab and was gone by the end of the year. To stay in a lab and advance, train others and be helpful and your boss will never want you to leave. Remember that your job, all jobs, is essentially to make your boss's life easier.

The diamond films can be reused until they are worn out, so the cost is amortized over several samples. I have seen single use labs and the opposite ones where the holes in the film had to be huge before the film was thrown out. As in most things, moderation is best. These films are about $30 each as of 2020, much costlier than the SiC grinding papers. If you have infinite budget then do what you like, but each dia-mond polishing film can be used five to ten times easily without compromising the samples being prepared; just use different sections of the film. Another consideration on the repeated use of films is the possibility of cross-contamination if you are polish-ing different materials from sample to sample. If you plan on doing any composi-tional analysis on the final cross-section use of fresh sheets is a good idea.

The typical process is to polish with 30-micron film (dark green) until you are close to the target and then move in with 15 and then 9. Going this aggressively may put cracks into the base die material. So, when cracking in the die is what you are looking for then start with a lower number paper or do ion milling only. The 6-micron film will still move the section forward but slowly and 3 and 1 basically do not move the section but are cleaning up scratches. The next steps are where the process varies quite a bit. The standard routine is to use a brown (final A) or black (chem pol) cloth with 6, 3, and then 1-micron colloidal diamond. Use a different cloth for each, otherwise the left-over larger particles will ruin your polish and your day. Finally, using a brown cloth; you can use 0.5-micron colloidal silica (white or blue is a personal preference with the variant being the pH of the solution). I use blue as it takes longer to crystal-ize and therefore is easier to clean up. When dealing with soft materials in the section like copper I use an extra step with Masterprep. It is best for removing the small scratches in the soft metals like copper. It is however sticky and a pain to clean off.

No matter which method is used you will want to ensure that you are using DI water in the process otherwise it is impossible to get a clean final product. Rinsing and drying for inspection after each step to ensure you are actually done with a step is highly recommended. The method of doing a step for a set time and just moving on will result in a lot of rework. To inspect a low magnification, 10-150x, stereoscope is best for the early stages and a higher power one, 50-2000x, for the

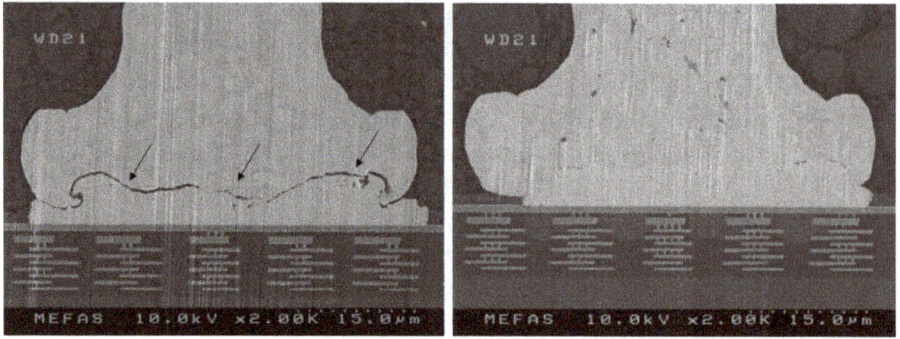

Figure 14.5: Polished cross sections of failing and good ball bonds. Note that the scratches are all in one direction on the failing part, but are not interfering with the analysis of the failure. The small black spots in the good ball bond are embedded particles from the polish, but are also not interfering with the analysis. Image by EAG.

final inspections, before SEM. Also, be sure to clean and dry the section before examination under the microscope as water will cover up a lot of scratches.

At each step before going to the next level of polish, the scratches should be checked. This is best done by polishing in just one direction for 30 seconds. All the remaining scratches should line up with the direction of polish. Examples of aligned polishing scratches are shown in Figure 14.5. If they do not you have some more to do at the current level.

After the final polish steps, it is very important that the surface be clean before images are acquired. This will require rinsing under DI water and using special soap with a soft Q-tip, cotton or foam will do, to remove the Materprep or colloidal silica. Residues, such as those shown in Figure 14.6, will make your section look terrible in the SEM.

Figure 14.6: A cross section that has residue from improper cleaning. Image by Dan Sullivan.

Then, thoroughly dry the sample. Usually, a filtered air gun will mostly dry it, but 15 min in a low heat oven or under heat lamps will do nicely. Wet samples in the coater or SEM will bubble up and move residues all over the sample and you will get to go back to the fun of final polish and clean again. An example SEM image of a well polished and cleaned sample is shown in Figure 14.7.

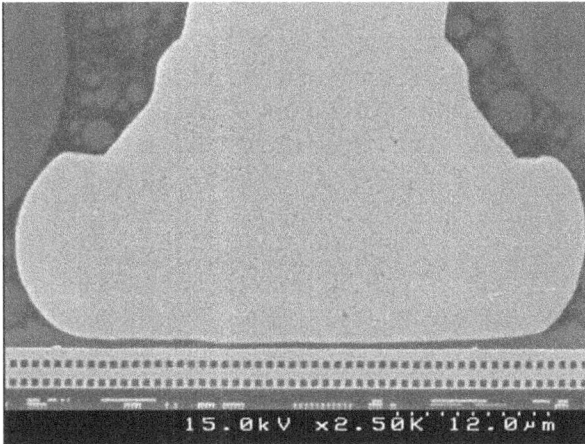

Figure 14.7: Extra final polishing of a good bond with all residual scratches and polishing media removed. Image by EAG.

Cross sections of packaged parts, or basically anything other than a die, requires a different approach. First, the sample may need to be reduced in size to allow potting in epoxy and polishing on a wheel. Most standard polishing wheels have 8-inch diameter wheels. Larger ones get expensive fast and are used for more specialized samples. Any sample that will be bigger than six inches will be exceedingly difficult and the ones over three inches are a challenge.

All that being said let's go through the typical 45 mm square device. First, decide where you want to section to and what is important to see once you are there. If you need to see the die intact and its attachment to the encapsulant and the die attach or underfill then you cannot cut it with a saw in the die area, we can cut next to the die without hitting it and we will be fine, so we can save a bit of potting epoxy, time grinding, and our arms. If, however, you want to look at the solder bump on row 18 and do not care about the die then you can cut to within two rows of the row of interest with a diamond saw and then pot the sample up. I do not like to use harsher saws unless the sample is very tough, metal or ceramic for instance, and there is no chance of ruining the area of interest.

To determine which SiC paper to start grinding the sample with initially will depend on two parameters: (a) how far away from the target area are you and (b) what can your sample take. With silicon die, you will shatter the die if you hit it with any grit harsher than 400, so do not start with 60 grit on the sample if you need the die intact once you are at the area of interest. If you are starting with the whole device and there is a cm between the edge and the die, use a diamond saw or harsh SiC paper to get 90% of the way to the die and then change over to the gentler paper. I have experienced that anything harsher than 400 grit will always crack a Si die.

The 400 and 600 grit paper will still move the section well, just more slowly than the harsher sheets. If there is a considerable distance to travel, say from the edge of the die to 15th row of solder bumps, the use of a semi-automated tool like a Tech Prep to do the long travel is very good. A maximum travel distance allowed can be set on the tool so you will not go through the area of interest. Just do not abandon the tool for too long as the paper can tear and the puck can fall off the tool. So, I recommend doing something else in the lab while the autopolisher is going and turn it off when you leave the lab.

Use the 600 grit paper to arrive just before the desired final location, this should be less than 100 microns from the area of interest. Once there and the scratches are gone (or all the same size and aligned) switch to 800 and then 1,200 grit. These steps usually take 30 s each and require switching the polishing direction 90 degrees away from the last orientation and then checking for scratches. Any scratches observed should all be in the same direction as the last polish before considering that step done. If no progress is being made, then switch to fresh SiC sheet. Worn out paper grinds much slower than a fresh one.

Do not go on from the SiC if there is still material to be removed, the line of polish is incorrect, or larger scratches exist in the area of interest. Polishing will not fix any of these in a human timeframe. If all the above are ok then proceed to the polishing steps.

The exact steps vary depending on the analyst's personal biases and the specifics of the sample, but I do the following for Si die in packaged devices: cut the extra substrate on the side I wish to polish on, and then pot the device in epoxy with the side I will polish to at the bottom of the cup. The next day when the epoxy is hard, I pull the sample and epoxy out of the cup, use a diamond saw to cut any substrate sticking out the top of the epoxy, and then cut the epoxy down to a size that is easy to handle and reduces the material I will have to grind and polish. I then grind (with 240 grit SiC paper) the sharp edges on the sample and remove any remaining substrate on the side of interest. This is also the time to make the far side flat and at the angle desired so the imaging in the microscope later will be easiest.

The polishing process is hard on your hands so wearing gloves is good if you can still have the control you want; getting rid of sharp edges of the epoxy is helpful too. Make sure your first aid kit has bandages.

Then, polish the die with 400 grit paper and proceed until close to the area of interest. During the process, I will check in the low-magnification microscope and the X-ray tool as needed as illustrated in Figure 14.2. In the grinding process, the angle that is desired for the end inspection may not match what you start with and you will have to make adjustments by applying pressure to the sides of the sample and turning it to have the faster grinding portion at the outer edge of the wheel. This is an acquired skill. When a new analyst is working on an important sample, I will have them get to the row of solder bumps next to the one of interest and then I will take over and show them the steps and have them do the inspections. They can learn the final steps on practice parts.

After the section is in the row adjacent to the area of interest and parallel to the desired final polish, I switch to 600 grit and proceed until I am almost in the center of the desired area but just shy. Then, I use 800 grit for 30 s and turn the part 90 degrees and do another 30 s. Next, I check for the scratches. If they all line up then I am done and move on. If there are some scratches that do not line up, we have to decide if they are worth removing. If they are not in or near the area of interest then ignore them and move on. If they are in the area of interest and not much larger than the other scratches that do align with the last angle of polish then do another round of 30 s in each direction, but be sure not to align exactly with the scratches you are trying to remove as this will exaggerate them.

If the scratches are much larger, then you will have to go back to a lower grit to remove it. If this will cause your area of interest to get removed, you will just have to live with that or take images with them. This is why it is important to get even polishing and check each stage before moving on to the next one. 1,200 grit is not the time you want to make major fixes.

The process above is repeated for 1,200 grit and then an orange cloth is used with 6-micron colloidal diamond. Some people use a red cloth but I like the orange as it is very flat and gives results with almost no rounding. Use a different cloth for 3 and then 1-micron diamond slurries. I write on the backs of the cloths with Sharpie and can then clean and store for reuse. Do not use the same cloth and just spray on the next slurry. This will cause you misery and waste a lot of your time. I clean the cloth (just rinse in DI water) after every use, but a few 6-micron boulders are going to be missed and stay on the cloth. These will delightfully put a huge scratch exactly in your area of interest just as you are finishing the last polish at 1 micron. So use the separate cloths, they are cheap.

After the end of the diamond polish, I switch to a chem pol (Allied High Tech), black cloth, to use 0.05-micron colloidal silica, the blue stuff. Once the sample looks the way I want, I can stop and go to inspection or if there is a fair amount of copper or the silicon still has some scratches that I do not like I use a last polish with Masterprep or an equivalent. This is a 0.05-micron colloidal silica but with a different chemistry that is very good at polishing out the scratches in copper and silicon. It is a bit sticky and requires more than the regular cleaning to get the sample in the desired state. I use a diluted soapy mixture with a cotton or foam swab to clean the surface and rinse for 30 s in DI water. Blowing the surface dry with clean dry air is a trick and low pressure and the right angle helps. If white residue covers that sample, then you must clean it again. It is very important to get the part clean and dry before preparing the part for SEM inspection. I once had to clean a part for an hour to get the area of interest right.

A word on the grinding and polishing papers/clothes. The SiC papers are cheap, maybe 50 cents a paper, so I recommend saving the ones that are very lightly used and then throwing away any that get worn. I had a technician that would use these until they fell apart. I lectured him on the costs per sheet and the cost of his time. A worn sheet is slower to grind than a new sheet and once it is 3x slower it should definitely go away. The polishing clothes are more expensive and should be cleaned; a good rinse should do and they should be saved and labeled on the back as to the solution to be used on them. Each cloth should only see one type of suspension. So, a collection of three orange polishing pads for the colloidal diamond, a brown (final A) or a black (chem pol) for the 0.05 Si blue (or white, but I hate the white) and one more for MasterPrep does it for me. I store these in a drawer with separators or a in box type file system to avoid cross contamination. Although it slightly adds to the initial investment, I find buying separate wheels for each polishing cloth helps if you plan to adhere them in place. The downside is that these then become shared tools and you are at the mercy of the sloppiest user.

Chapter 15
Parallel lapping (p-lap)

P-lap is also referred to as delayering and can apply to die, package substrates, and PCBs. The basic process is to inspect the current layer and then remove it in a controlled manner and then inspect the next layer; usually by optical microscopy or SEM. This is commonly done once a defect has been localized by another method: EMMI/OBIRCH/IR or designated area for inspection.

For die, the process involves lapping with abrasive materials (SiC and Diamond impregnated sheets, or colloidal slurries on polishing clothes) just like in cross-sectional work. In this case, we are trying to remove just microns of material so the harsher grade SiC papers are not used. Typical p-lap will use a brown cloth with 0.05-micron colloidal silica, polishing for a minute, rinsing with DI water, checking under an optical scope, and then repeating the process until the area of interest is cleared. This requires a personality that can repeat the step carefully many times.

Plasma and wet etches can be used to remove the oxide and metal layers respectively. The use of these techniques allows a larger area of interest to be maintained as flat compared to just polishing. The polishing almost always removes the material from the edge of the die and corners faster, so it is difficult to have areas of interest maintained through the whole delayering process that are more than 100 microns in diameter just by polishing.

There is a special case for suspected ESD-damaged samples when there is more than one sample or the confidence in the issue being ESD damage at the active region is very high and that is a hydrofluoric acid (HF) strip. In this case, silicon die only, the oxide layers are etched away with HF and this carries all the metal layers away once the oxides are etched. Often, this can leave a dirty surface which requires some cleaning by ultrasonic IPA rinsing. The part can then go straight into inspection by SEM after the HF and clean-up is done. This is why it is often advisable to polish several layers off the die before doing the HF dip as this reduces the materials etched and the resultant residues. This method is quicker and therefore less costly than a full p-lap, but it only allows inspection at the silicon and does not allow inspection of any of the metal layers. An example of an HF stripped exposing ESD damage is shown in Figure 15.1.

For the die it is important to be able to identify the area of interest as the layers are removed. Sometimes this is possible with the use of GDSII files, but not always. Sometimes, it is necessary to laser or FIB mark the die prior to starting the delayering process. This involves making one or more holes in the die near the area of interest. These need to go all the way through the oxide layers so the mark will still be present when the delayering is complete. These marks are usually a couple of microns square or bigger and far enough away from the area of interest that no induced damage will be seen in the area of interest. Generally, 5–10 microns is a good distance. One spot has the issue of losing the orientation on the sample as can two spots. I use three

https://doi.org/10.1515/9781501524790-015

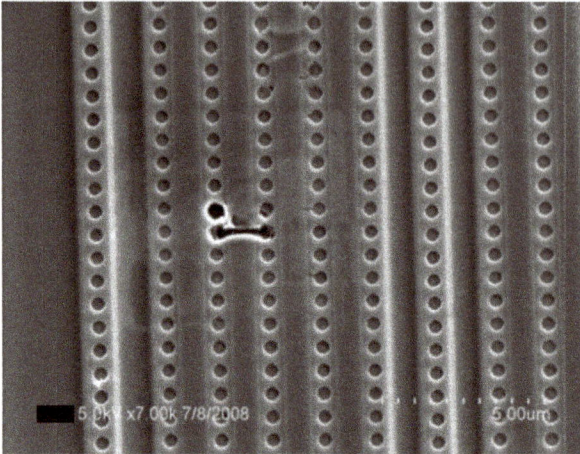

Figure 15.1: SEM image of an ESD-damaged sample after an HF strip. Image by EAG.

spots as the area of interest is then well defined as shown in Figure 15.2. I call it idiot proofing as after making the mistake of using only one spot once and having to spend a lot more time in the SEM than I should have needed. I do not want to be an idiot again.

Figure 15.2: Three laser marking spots on die, denoting the area of interest in the middle. Image by EAG.

Initial images of the area of interest are very helpful for later use in reports and presentations to verify that the correct area was examined and low to high magnification shots allow everyone to agree on what was done and where. I recommend when dealing with high profile cases or difficult customers that a set of images before any polishing is done be sent to the requestor stating that this is the area to be analyzed and request verification. The images should start at low magnification so the device orientation is clear and the area on the device is identified. Steps to higher magnification should allow easy determination of where on the device the analysis is to be done. A picture of New York and the next image being a close up of a penny on a sidewalk is not the way to find the penny. Intermediate images are required.

When I first started my FA career, the semiconductor devices had Al metal traces in the die. These were great as they had different materials in the via stacks which acted as etch stops. This allowed for a method of delayering involving plasma etch to remove SiO_2 to expose the top layer metal for inspection and then a wet etch to remove the Al traces. This allowed inspection of the via tops and then repetition of the process down to M1. Many modern devices now have copper traces which attach directly to copper vias, so there is no etch stop. Timing etches can be done to just etch one layer at a time and we used this in some of the initial FAs on such devices. As you can probably guess, this is not a great method. This is why we use polishing of the metal layers with plasma removal of the glass layers as the norm now.

For package substrates and PCB boards, the process is very similar to that used on die, but the polishing material changes. Going a couple microns at a time with 0.05-micron slurry to delayer a PCB would take a very long time. To quickly remove the PCB or substrate to free a device or die is common and 60 or 120 grit paper can be used to get most of the way through very quickly. Only when the die or solder balls are very close, you switch over to 240 or 400 grit to finish up the removal of the desired material. In this way, a die can be removed from a package, or a part removed from a PCB, without the use of heat. The process is a bit of an art form and I recommend a practice part or a fair amount of experience before doing this on a part that matters.

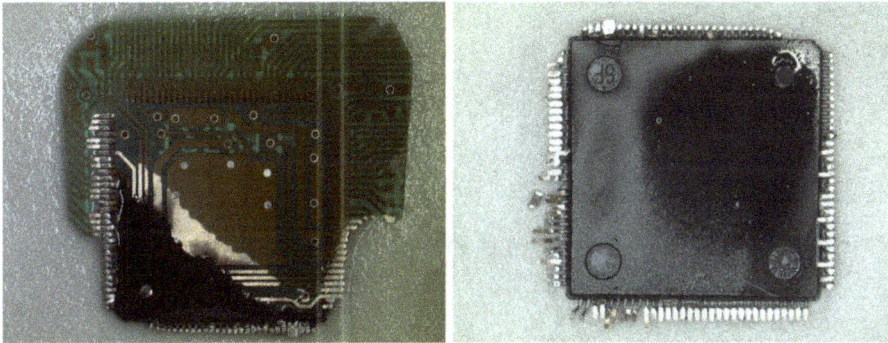

Figure 15.3: Device mounted on PCB after removal of the PCB by polishing (partial and full removal of the PCB). Image by EAG.

If, however, the goal is to inspect the layers in the substrate or PCB, then a slower and more controlled method should be used. I have tried automated methods and never have I had success in getting the layers fully exposed and flat so that no portion is accidently removed. So, grinding/polishing by hand is the preferred method. Often, we are checking critical dimensions or looking for signs of a zap or embedded material causing leakage or a short. The traces and vias are exposed as the layers are removed and a check for the leakage/short can be made with either a probe station or multimeter and some sharp probes at each layer. Opens can of course be checked for in the same way. I have probed the top and bottom of a via stack to show there is high resistance or an open as a verification when a crack is partial or there is conflict agreeing on the image showing a crack. Those who want a different answer will try to claim the alleged crack is just a scratch in the metal. The electrical measurement negates this argument.

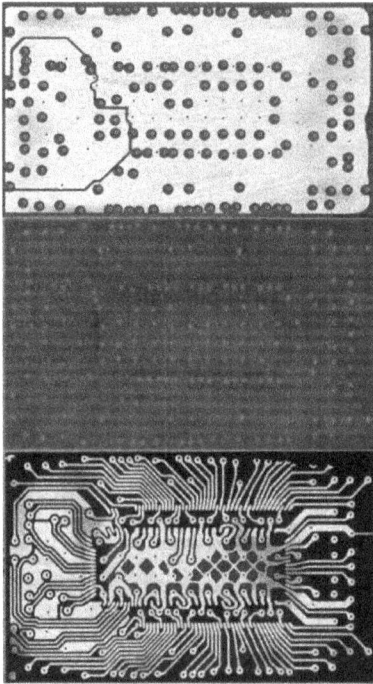

Figure 15.4: Lapped substrate: layer 3-2 traces, layer 2-1 contacts, layer 1 traces. Image by EAG.

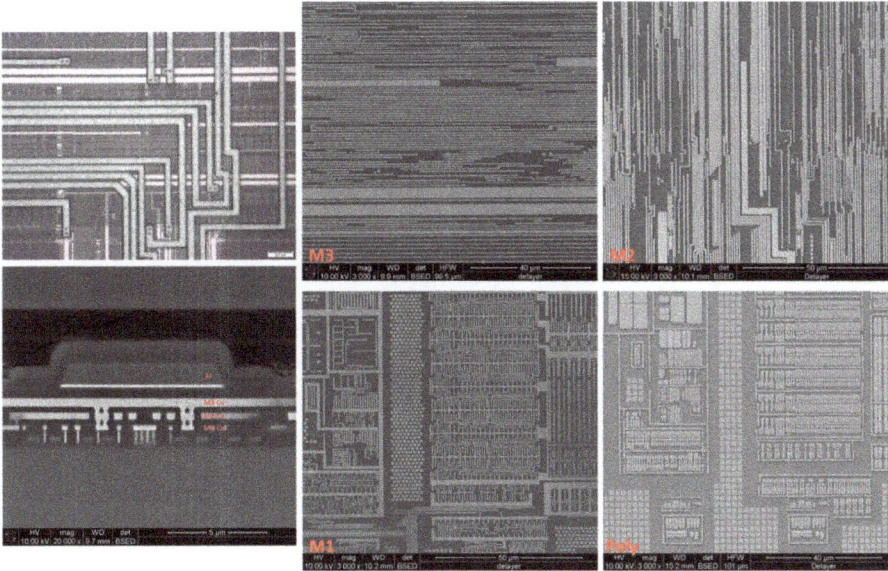

Figure 15.5: Checking each layer in a Cu trace device. Cross section, optical image before delayering includes RDLM3, M2, M1, and poly. This is then followed by SEM of each metal layer as it is exposed. Image by EAG.

Chapter 16
Focused Ion Beam (FIB)

There are several types of FIB. Single-beam FIB does the cutting and viewing with the ion beam and generally has lower image resolution and is not usually used for FA. Circuit edit (CE) FIB is used to perform surgery: cut and jump lines in the device to check possible alterations of a device quickly before doing a re-spin. Re-spins are costly and can take many weeks to get new samples back out of the fab, so CE FIB is used to check any desired alterations in a day or even hours before going back to the Fab.

In the FA area, the CE FIB is used to put down markers (instead of using a laser on a substrate as was discussed in the previous chapter) or to place probe points down into the circuitry to allow those lines to be checked electrically via a probe station. Alterations can also be made which can be checked on the probe station or on a full ATE.

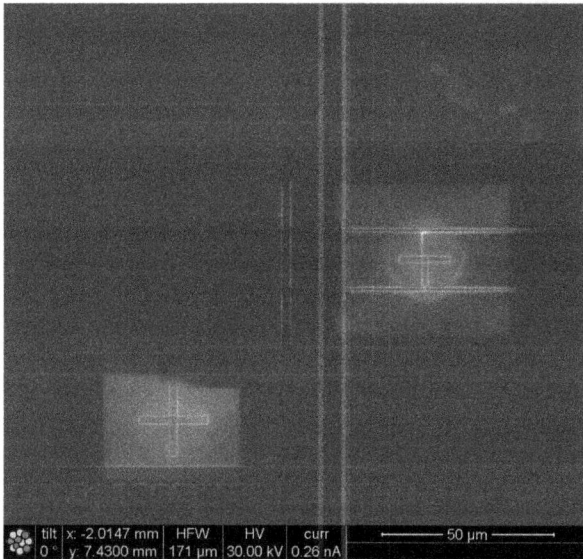

Figure 16.1: Two probe points are created on die with FIB for electrical checks. The cross patterns are used to help the probe tips to land and stay in the corners of the cross. Image by EAG.

https://doi.org/10.1515/9781501524790-016

Most often used in FA work is dual beam (DB) FIB. This tool has the ion beam and an off-angle SEM system on one tool. When the tool is aligned and the sample is placed at a particular height, the SEM beam and the ion beam are coincident. This allows a trench, really a ramp, to be dug by the ion beam and the electron beam of the SEM can image the wall at the end of the ramp. This is often referred to as cut and look process. In this way, a defect spot localized by a different technique can be examined with very small and controlled steps being made by ion milling the ramp at steps as small as 0.1 micron with SEM images and if the tool is equipped with an EDS tool, then elemental maps or spot spectra may also be obtained at each step. The normal practice is to start a large trench, 20-microns wide by 5-microns deep is typical, outside or at the edge of the defect location and slowly polish into and through the defect area taking multiple images and EDS data as desired. In this way a stop action movie of the defect can be obtained.

Figure 16.2: DB FIB sectioning and imaging showing a buried defect in a Si device. Image by EAG.

The DB FIB is also nice in that when a defect is found and higher resolution imaging is needed the sample can then be prepared for TEM directly from the DB FIB process. The SEM systems, including DB FIB, have a resolution in the teens of nm while TEM is usually ~0.1 nm.

Figure 16.3: FIB sectioning to make a TEM lamella for higher resolution imaging of defect regions. Image by EAG.

Some DB FIB tools are equipped for cryogenic operations. This allows the sample to be cooled during the operation. This is important for some materials like Li-ion batteries and delicate materials that the energy of the beam would melt or alter without the cooling. These tools are expensive and require long stabilization periods to arrive at the low temperatures. An alternative is a cryo-cooled plasma FIB (Cryo-FIB). This tool etches faster by a factor of 6+ and is capable of creating much larger sections.

Chapter 17
Scanning Electron Microscopy (SEM)

SEMs are wide spread in analytical labs and on universities. It is often the first large-cost tool general labs acquire. The resolution, maximum magnification, allowed sample size, and image quality varies a great deal depending on the make, model, vintage, and upkeep of the tool and on the abilities of the analyst.

The tool used will depend on the need, the availability of tools, and the costs. If you are concerned with particles or cracks in a sample that are hundreds of microns in size, the sample is small (less than an inch across) and the image does not have to be sharp, then any instrument will likely do and almost any analyst.

The more the image quality requirements increase, the more the appropriate tool selection will tighten, and the more skilled your analyst will need to be. The standard emission sources are: tungsten which is an older and lower resolution technology, cold cathode, and field emission which is the newest and highest resolution system.

Table top systems have become more widespread as they are significantly less costly than full size systems. They typically can only reach 20,000x with reasonable image sharpness. This is often good enough for many applications such as substrate evaluation or checking solder attach.

Mid-range systems can usually achieve 50,000 to 150,000x with reasonable image sharpness. These are used to examine a wide variety of samples including semiconductor devices (both x-section and p-lap).

The newest high-end systems can get to magnifications as high as 700,000x and require special setups in their locations to remove vibration, electrical fields, and acoustic interferences. These of course require a better class of analyst and much more precise tweaking of the beam to get the best images, but they are spectacular.

Typical uses of SEMs in FA work include: imaging at magnification beyond the ~2,500x maximum for optical systems, large depth of field requirements, combination use with EDS to overlay the SEM image with EDX point analyses or elemental maps, and passive voltage contrast (PVC) to determine open or short locations in p-lap samples. Layer thickness measurement is also frequently used to determine if a device layer structure is compliant and may be contributing to an observed failure.

https://doi.org/10.1515/9781501524790-017

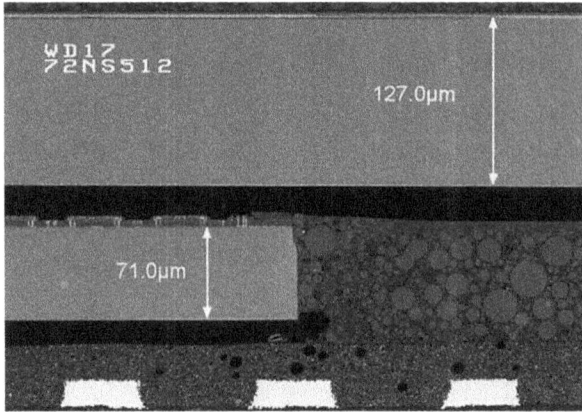

Figure 17.1: Measurement of layer thickness with an SEM. Image by EAG.

The disadvantages of SEMs are: the images are gray scale and not in color (although false color can be applied in post processing), the images are in line of sight to the beam and detector and only of the surface (the beam energy has some effect here but the image is primarily of the surface with very little if any depth information), and insulating samples will charge and interfere with the imaging (this can be dealt with by coating the sample with a thin film of conductor (most commonly Au, AuPd, Ir, Ni, or C) or by utilizing a variable pressure SEM that allows a small amount of gas into the chamber to dissipate the charge.

Without a coating to dissipate the buildup of charge from the electron beam, work can only be done at low magnification and usually only in back scatter mode (BSE). Otherwise, the image will bend and warp and then the image will flash and good images will be impossible to obtain. Once the charge is built up it stays and the only way forward is to remove it from the chamber to air and let it discharge.

The deposited thin dissipative coating is very good at preventing the buildup of charge, as long as the surface of the sample is grounded. The grounding is through the sample holder and the stage, but the face of the sample must be put in electrical contact to the holder. This is often done with carbon or silver paint: these should be thoroughly dried before being placed into the chamber otherwise they may bubble and get onto the area of interest. This is very unpleasant as the material needs to be polished off and the sample must be recoated, the paint reapplied and dried. Also, your pump down will be much longer with wet paints. Alternatively, metallic tapes, Al and Cu are common, or spring clips (CuBe or SS) can be used to make electrical contact with the sample face.

Using an SEM to get reasonable images is typically easy for trained personnel on a well-maintained tool; however, getting the best images possible from a tool is a mix of science and art. I can still jump on a tool and get reasonable images in 10 min if the beam needs a little alignment and the stigmators are not perfect.

I have people in my lab that can then sit down and in another 5 min get the images to be sharper and clearer in a way that I just cannot do. Experience and practice are the keys to being a good SEM analyst, but some natural talent is involved.

That being said, there is a wide variety of SEM systems. The field emission (FE) systems are much better than the Lab6 filament systems and well-maintained systems are always better than their equivalent with less rigorous maintenance. The desktop tools are fairly good to about 15,000x, while old Lab6 tools can sometimes get up to 30,000x before their images get blurry. An FE SEM should be able to take good images up about 100,000x and the newer tools can do several 100,000x. This is very impressive until we go to the next chapter and see what a TEM can do.

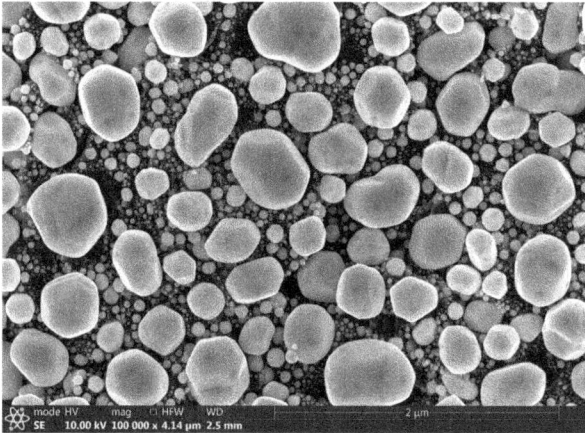

Figure 17.2: SEM image of Au particles on C at 100,000x. Image by EAG.

Chapter 18
Transmission Electron Microscopy (TEM)

For TEM, the image is generated by passing an electron beam through a sample instead of just hitting a sample and looking at the electrons that come off the sample as in SEM. The technique also uses beam energies much higher than those in SEM; 200–300 keV as opposed to 30 keV and below. This results in much higher spacial resolution, on the order of atomic spacing. TEM samples were previously made by mechanical polishing but now are almost exclusively made using FIB tools. The FIB is used to cut to a specific location, often a defect, and then thinning the sample from both sides until a lamella less than 100 nm is created. This is then removed from the sample and attached to a grid for handling and placement into the TEM instrument. The thinner the piece, the better the images, but the thinning process has risks and so there is a tradeoff.

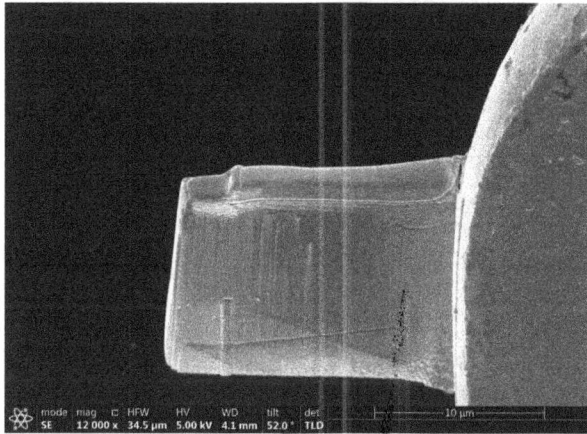

Figure 18.1: TEM lamella. Image by EAG.

Fancier and more expensive additions to a TEM include scanning TEM (STEM) which utilizes an off-angle beam incidence for imaging rather than the perpendicular beam incidence used in standard TEM. Additionally, imaging can be improved with integrated differential phase contrast (iDPC-STEM).

TEM instruments can be purely for imaging or commonly additional capabilities are added onto the system like EDX (energy dispersive X-ray spectroscopy) and EELS (electron energy loss spectroscopy) for elemental identification of spots or mapping areas. EELS can also provide some chemical information as well.

https://doi.org/10.1515/9781501524790-018

Figure 18.2: Direct imaging of Ga and N atomic rows in GaN by iDPC-STEM. Image by EAG.

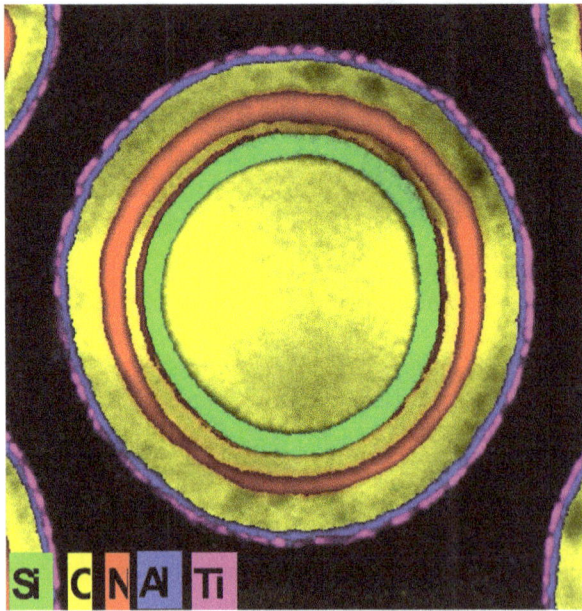

Figure 18.3: EELS elemental identification of a 3D NAND cell. Image by EAG.

EDX and EELS both provide elemental composition, except for H and He, with detection limits of 0.1 to 1 atomic percent depending on the element. In SEM, this is from a depth of several microns as the beam spreads inside the sample in the shape of a tear drop. So, at 20 kV, the beam penetrates roughly 5 microns and is about 2 microns wide at the middle. The elemental composition is from this entire area, so very thin films may be missed entirely. In TEM, by contrast, there is only 100 nm or less of sample and the spreading electron path effect in SEM is not present here and thusly the spatial resolution is greatly improved.

Overlaying the maps generated by EDX and EELS onto the TEM images makes clear the material's layout and the location of any contamination.

There are new improvements to the TEM systems including aberration correction and image processing capabilities that allow for improvement of the resolution of images and that also enable stress measurements at interfaces. These are of great interest in cutting-edge semiconductors.

The use of TEM in FA work is usually after a fault localization by EMMI/OBIRCH or IR and then DB FIB with the defect being too small to get all the details desired in the FIB, so then a lamella is made and examined in the TEM.

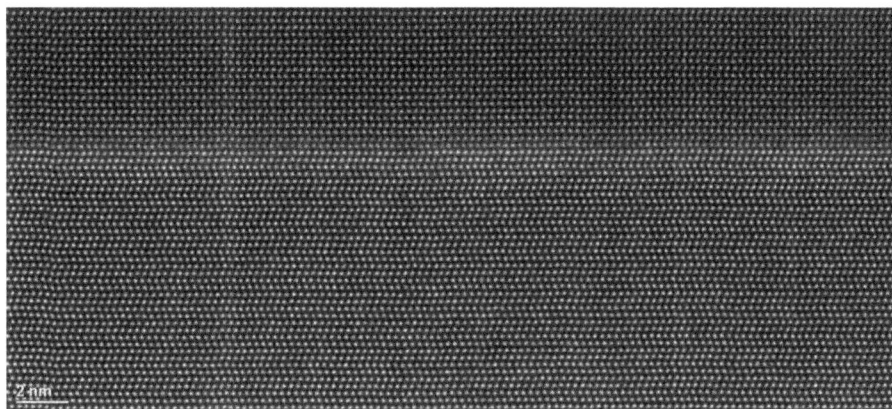

Figure 18.4: TEM interface of an AlN layer on 4H-SiC with atomic resolution. Image by EAG.

We are very lucky to use the modern FIB tools to prepare the vast majority of TEM samples. In olden days, these were done by hand polishing or using dimplers to thin the TEM samples. Both required the patience of an angle and skill. So, I did not do this myself.

I will never forget the moments in the lab when someone would call out, "no one move" as a TEM sample had just sprung into the air and onto the floor. This was usually followed by many minutes of intense inspection of the floor with flashlights while on all fours.

Chapter 19
Energy Dispersive X-ray spectroscopy (EDX)

EDX is also referred to often as energy dispersive spectroscopy (EDS). It is a tool attached to many SEM instruments and some TEM systems. If you have an SEM and for 10% of the cost of the SEM itself you do not add an EDS tool, you either have a very specific measurement you need and nothing else or you are a fool.

The EDX process occurs in an SEM when high-energy (up to 30 keV) electrons strike a sample. This creates the electrons that make up the SEM image and also produces X-rays. The X-rays are dependent on the elements struck by the beam and have energies independent of the beam energy; provided the beam energy is approximately 30% higher than the X-ray produced to get the process to go. When the X-ray process occurs, it is due to the relaxation of electrons from higher energy shells into the vacancies in lower ones that were created by the incident beam. The transition energy from the higher shell to the lower one is what determines the energy of the X-ray. Elements with many electrons have several such transitions. Each element has a set of electrons in energy shells: for instance, Fe has 1S, 2S, 2p, 3s, 3p, 3d, and 4s all full. Each of these shells may lose an electron due to the electron beam and then an electron in a higher energy shell may relax and fill the empty slot. This releases energy equal to the difference in the two shells in the form of an X-ray characteristic of the element and transition. This emitted X-ray is then detected by an X-ray detector with a path collimated to be coincident with the electron beam, in most systems this is typically at 10 to 15 mm below the SEM pole piece. The shells are referred to by letters (KLM) and the transitions called K Alpha or K Beta for example denoting the energy shells involved.

While there are some overlaps of peaks in EDX, there are not too many. The worst is W and Si with a couple of others occasionally cropping up. More modern detectors with good software can model the entire spectrum and sort out these overlaps except when one element is present in high concentration and the other in very low concentrations. In these cases, there are other techniques to determine the presence of the suspected lower concentration material.

In using EDX, it is helpful to define what you are trying to determine before the analysis as this helps determine what beam energies should be employed. The lower the beam energy, the shallower depth the electrons penetrate into the material and thusly where the X-rays originate from, this is due to the electron penetration and not the X-rays themselves. At 20 keV, the X-rays come from a region from the surface to approximately 5 microns deep while at 10 keV this is ~3 microns and at 5 keV ~1.5 microns. So, using this information and collecting spectra with multiple beam energies it is possible to determine if a material is uniform in the area analyzed or if one element is present at the surface versus the bulk.

https://doi.org/10.1515/9781501524790-019

There is of course a catch. To excite a transition that produces X-rays, the beam energy must be ~30% higher than the energy of the transition. Many metals have their distinct transitions energies above 5 and some above 10 keV. So, if these elements are of interest, then the higher energy beams must be used. This why standard settings are 20 or 30 keV with a 10 keV follow up spectra collected if the surface versus bulk elemental determination is of interest.

Using multiple beam energies require more time to acquire more spectra obviously, but it also requires the beam energies used be calibrated so the SEM is still focused on the same location. In more modern instruments this is less of a hassle as the tool can be programmed to hold parameters for multiple beam voltages.

To make maps with EDX, it is necessary to select the windows of energy to be collected before the map is generated. This is why a full spectrum is often obtained beforehand to see which elements are present. Even after the data has been collected, it is strongly advised that a knowledgeable person review the individual element windows and set the integration areas before the maps are rendered. I have seen several instances when noise was counted as real signal. The hysteria around nonexistent contamination being found is bad and then having to explain what happened is no fun either. From that point on, all data from the lab will be suspect for quite some time.

On several occasions, I have had to review and correct the endpoints for collected spectra and in one case it was necessary to repeat the experiments. In this case, the analyst did not save the raw data after making the maps so review and correction was not possible. The imaginary Pb they reported on the maps caused much panic and hand wringing. The collection of the new maps with competent review showed noise was present and not the feared contaminant Pb. The extra day and the unnecessary stress went away after that and a rule was introduced that the raw data need to always be saved and when something bad was found I had to do a review of the data before it was released.

It is important to call out that competent chemists or well-trained spectroscopists either run the tool or be consulted when questions arise. I have seen a report that claimed that the major constituent of a solid sample in an FA was argon. After laughing, I called on the report writer and we reassigned the peak. Argon can be implanted in materials and I have detected it with XPS at ~1%, but at room temperature it is not going to be the major constituent of a solid at STP let alone in the vacuum chamber of an XPS instrument.

A common use of EDS is to identify a particle or a discoloration. Here, some care must be used in the interpretation of the results and consideration given to the use of other techniques. For very thin, less than 1 micron, particles or stains, the EDX will be getting most of its signal from the substrate and not the material of interest. In these cases, two spectra, one at 10 and the other at 5 kV for example will help elucidate what elements are present on the surface versus the substrate. It is possible for very thin particles and stains that 90% or more of the signal will be

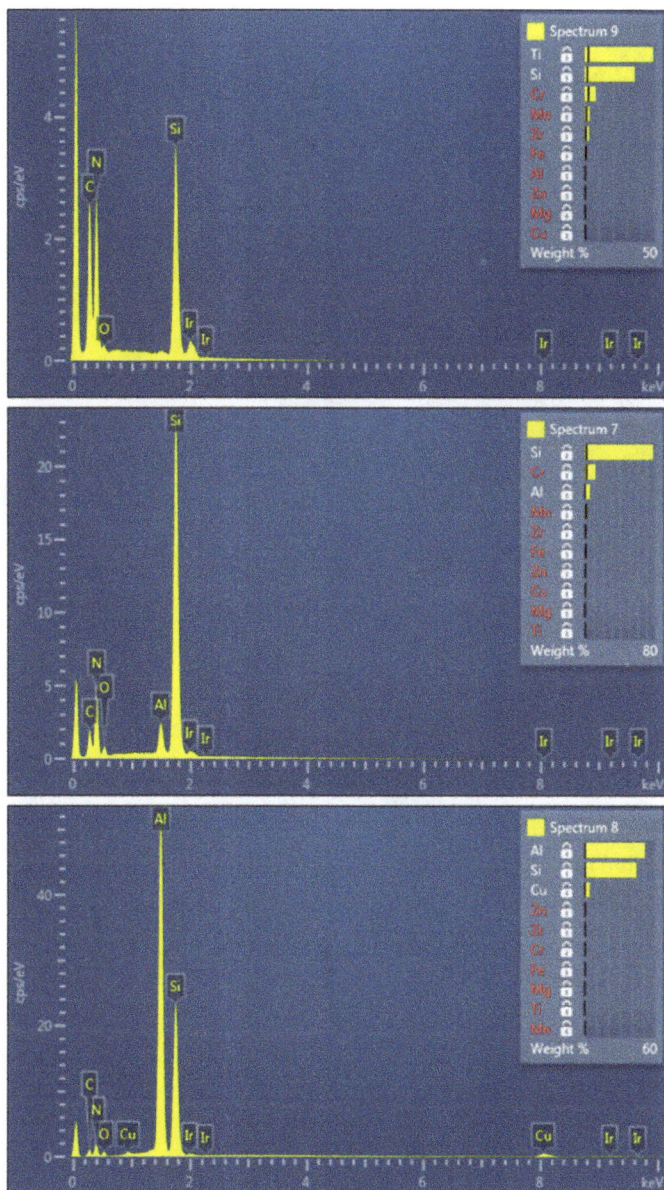

Figure 19.1: EDX spectra at 5, 10, and 20 keV. Note the increase in the C and O signals at 5 compared to 10 and 20 keV as the C and O are more prevalent on the surface versus the bulk of the sample. Also, the higher energy peaks are not present in the 5 keV spectra as they require more energy to be excited. Image by EAG.

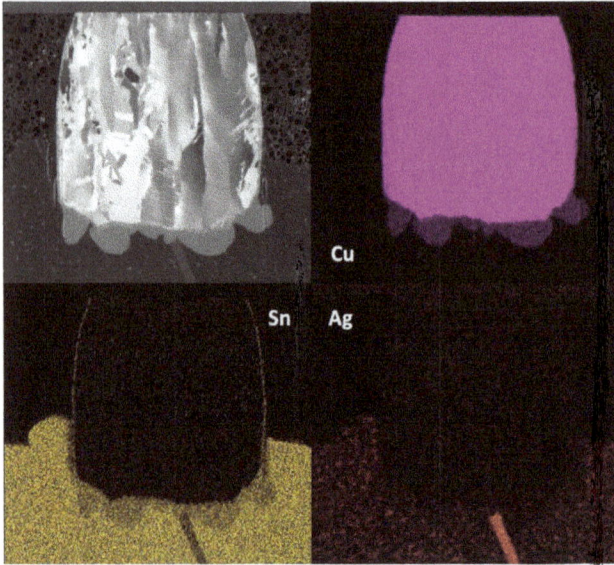

Figure 19.2: SEM overview with element distribution EDX maps of Cu, Sn, and Ag. Image by EAG.

from the substrate and in these cases alternative analytical tools should be used. X-ray photoelectron spectroscopy (XPS) and Auger (AES) are good candidates for thin film, particle, or stain identification work as they analyze only the top ~10 nm of a surface.

EDX is at best a semiquantitative tool. All concentration results from EDX should be viewed with large error bars. Absolute concentrations are not reliable but comparisons between samples run on the same tool under the same conditions are comparable qualitatively. A has more carbon than B for example.

Chapter 20
Use of additional materials analytical techniques and processes

There are many analytical techniques not mentioned in this book. The vast majority of FA issues in semiconductor and electronic devices are done with the methods described previously. However, when needed other techniques can complete or verify the results of an FA that is inconclusive.

There are several techniques I use most often to complete FA analyses. I describe each here and when they are applied and their limitations.

Auger electron spectroscopy (AES) provides elemental composition of very thin discoloration or small particles and to look for thin contamination layers at the interfaces of thin film stacks. AES is very surface sensitive, top 10 nm, and provides elemental composition, excluding H and He, with detection limits similar to EDX. AES does not

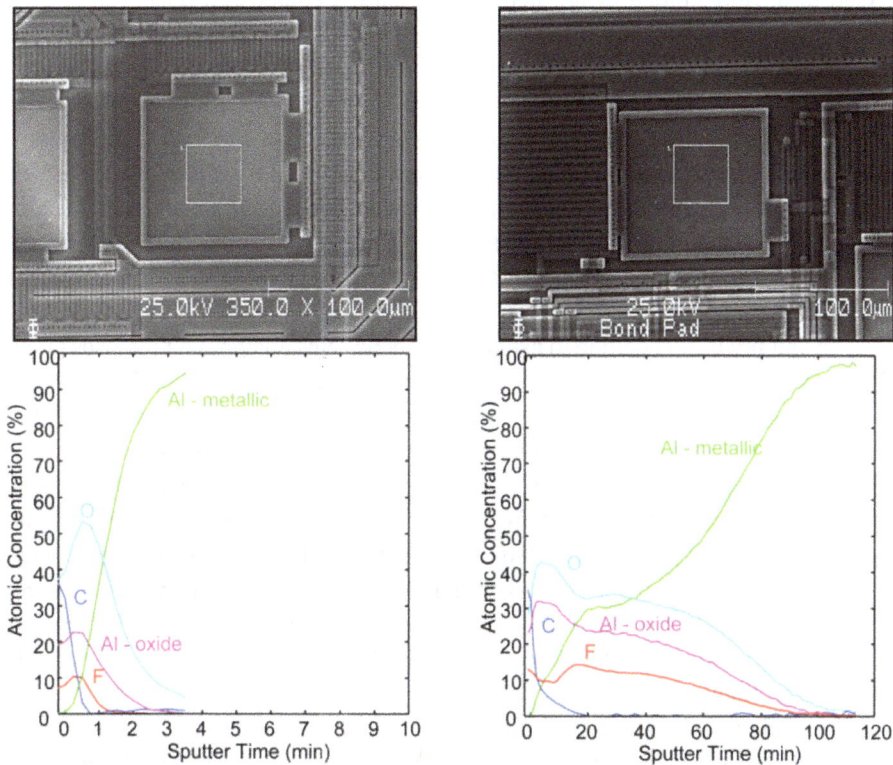

Figure 20.1: AES depth profiles of a good bond pad (left, oxide layer ~4nm) and a contaminated bond pad (right, oxide layer ~160nm). Image by EAG.

https://doi.org/10.1515/9781501524790-020

work on insulating materials and can be used for depth profiles but only to ~1 micron depth. The most common FA use of AES that I have seen is checking bond pads and the bottom of the lifted ball bond when the attach fails. Auger is great at determining if there is contamination or a thick oxide layer present that interferes with good bonding.

X-ray photoelectron spectroscopy (XPS), also called ESCA for electron spectroscopy for chemical analysis, is very similar to AES but uses X-rays to start the process. It has the same detection limits and surface sensitivity, but works well on insulators [2]. The minimum area analyzed is significantly larger for XPS with a minimum of ~15 microns but normally the area has 150+ microns. XPS does provide some chemical information, oxidation states, that AES does not provide. So, if you have C, Al, F, O, and Fe detected, you may be a chemist and know what is bound to what, but XPS provides measured values to show which elements are bound to which. XPS can be used for depth profiles much like AES.

Fourier transform infrared (FTIR) is commonly used when an organic contamination is found in the course of an FA on the sample or if the material itself is suspected to have degraded or not being the specified composition. FTIR uses light absorption versus wavelength to produce a spectrum of the material. This spectrum is a finger print of the functional groups present. In FTIR, a spectrum is obtained for a material that can be compared to the spectrum of a known material or more commonly the unknown material's spectrum is run in a comparison to a library of spectra containing many materials and the best match(es) are reported. The degree of match is also reported, chi-square value, so the quality of the match can be considered. FTIR also enables a fresh material to be compared to one that has been exposed to a harsh environment, such as heat, UV, or plasma, to see how the material is altered or degrades.

Dye and pry is a technique to examine hermetic seals and attach processes [3–4]. The most common use is to examine solder attach of devices to PCBs. The PCB may need to be cut to size so it will fit in the various chambers and fixtures used in the process. The purpose is to check the solder attach of many solder balls at once for cracking and nonbonding. The device and PCB are cleaned to remove debris and anything that may be blocking the ink from ingression to all the solder connections. The part is then submerged in red or UV ink in a vacuum or pressure vessel to ensure the ink gets in everywhere it can and air does not block any paths. The part is then removed from the ink and thoroughly dried. The PCB is then secured in a fixture and the device pulled from the PCB. This can be done in several ways with some methods clearly better than others.

The crude way is to pry the part off with a screw driver or similar object. This often results in some damage to the part where it comes in contact with the screw driver and since the force is not even or straight up the device can be warped and sometimes the ink can be smeared. A much better solution is to secure the PCB down and attach with glue or cement an eye bolt to the center of the device. The

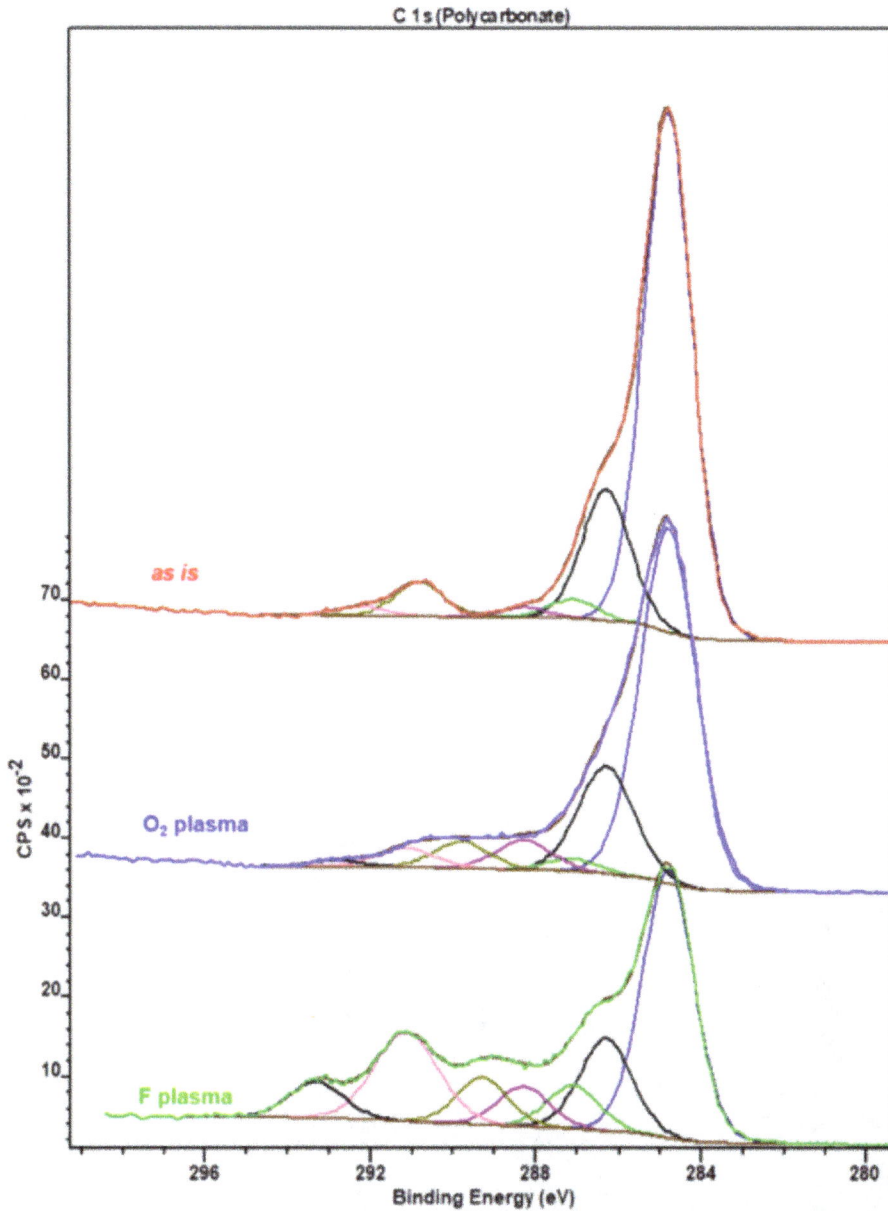

Figure 20.2: An XPS high resolution C 1S spectrum showing various C bonding: C–C, C–O, and C–F on a polymer after exposure to F or O plasmas. Image by EAG.

eye bolt is used to attach a line and a pull apparatus is used to pull directly up to remove the device from the PCB. Something like an Instron tool can be used for this with a bonus that the force required can be measured.

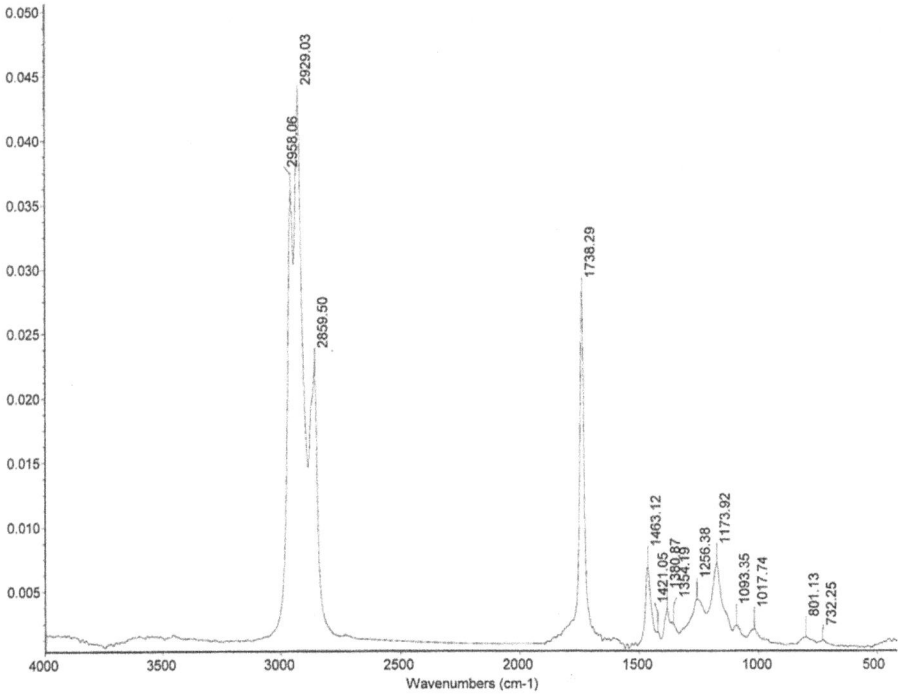

Figure 20.3: FTIR spectra of residue left in a process gas canister showing contaminants that ultimately resulted in multiple failures further down the line. Image by Arrhenius.

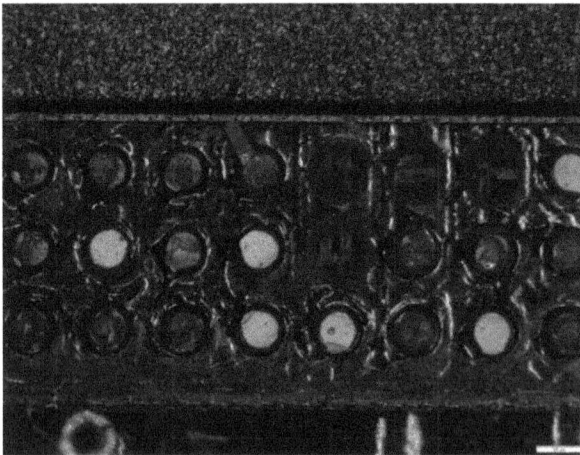

Figure 20.4: Image showing cracks and nonbonding present in solder balls through dye and pry analysis. Red indicates the cracking or nonbonding of the attach as the dye was able to cover the surface. Image by EAG.

The two sides now exposed are inspected and imaged with a grade being assigned to each solder or attach point. This is covered in a paper by Dell [3-4].

The use of red dye makes inspection easy unless your sample is also red. UV dye is also used but this requires the addition of a UV lamp and in my experience is more difficult.

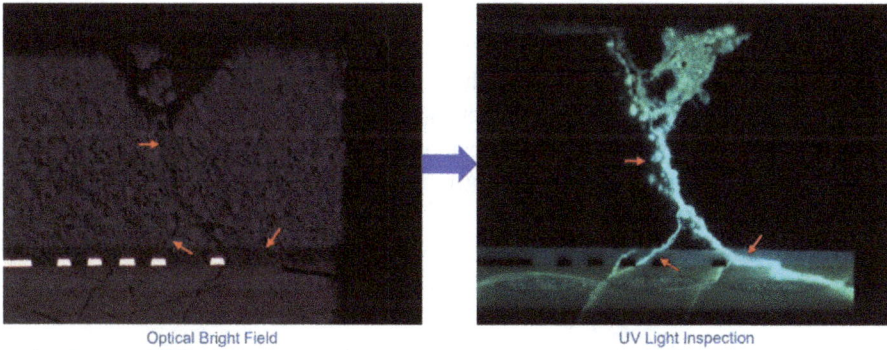

Optical Bright Field   UV Light Inspection

Figure 20.5: Image of dye penetrant test showing the crack in a cross section. Image EAG.

This process is also called dye penetrant when used to look for cracking in materials during a polishing or sectioning operation. If a crack is found and it contains ink, the material is discolored by the ink then the crack was present as received and not induced by the polishing process.

Laser decap or laser ablation. When this came out as a new process, it solved an issue for me right away. We were experiencing intermittent opens on our new packaged devices. X-ray was not great in the 1990s but we could see just the hint of a crack in the traces in the substrate. This always occurred at a turn in the traces and where the substrate had a change in thickness. We did some cross sections to show the crack, but the packaging group argued that we were inducing the very thin crack observed. So, we used parallel lapping to show the trace and the crack, but again the packaging group argued the crack was induced by the p-lap process. So, we used a very light sandblasting using wheat germ as the abrasive to expose the trace. More crying that we could be inducing the cracks and that we cannot be sure. So, when the laser system showed up in the following week, we exposed many, many traces and only the spot where we detected the crack in the x-ray had any kind of damage. The packaging group saw the images and gave up and accepted it was their issue.

This is an all-too-common occurrence when the group with the issue delays, cries and carries on which just delays the real work of fixing an issue. In this case, they had actually suspected they may have an issue when they had designed the traces to do a step up at the corner which was not a normal pattern. They wound up altering the layout and the issue went away. The laser system probably saved me several weeks of working on a nonsense issue. The proof of an image with the defect plainly visible is very difficult to argue against.

This tool is also excellent for pre decapping. That involves making a cavity in the epoxy in a packaged device above the die which helps decap (exposing the die) by wet etching the remaining encapsulant material. The precavitation allows less acid to be used as a portion of the encapsulant is removed by the laser oblation and this increases the success rate as the cavity holds the acid in place so it does not wander off and attack the substrate. It is important not to hit the die with the laser as this will damage the die. We have done tests and it always damages the die at the power levels needed to remove the encapsulant. I did do some work on this and we were able to clear the residual encapsulation after a decap in one corner of a die without altering the die. It took only 8 h. So, this was dropped as it is not on a human timescale for decap process.

This process can also be used on PCBs and substrates to allow probing or breaking of traces to change the layout in a crude way.

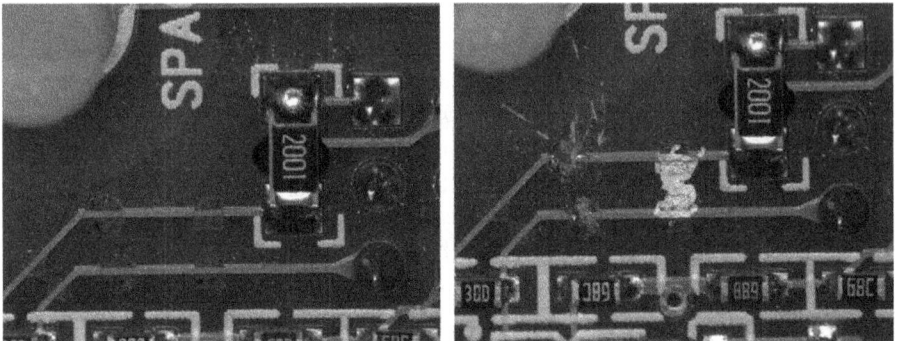

Figure 20.6: Optical image of traces exposed by laser ablations tool and with the traces cut and silver painted to reroute path for further testing. I call this caveman circuit edit. Image by Dan Sullivan.

For one customer, I repaired 40 units by exposing traces on a package and breaking two traces and using silver paint to connect the traces in a different way. These altered devices they used to confirm the fix would do what they wanted (in this way, it is a caveman version of circuit edit) and allowed their customer to start debugging their system with these devices until new packages could be made. This saved them a couple of weeks in the roll out of the systems.

This is by no means an exhaustive list of analytical techniques available. If you need a problem solved, do some research: online or through people in the field to find what you need. There are several independent service labs like the ones we currently work at that are happy to discuss your issues with you and make recommendations on analyses and the results that can be obtained. Just be sure to get a quote before you start any work with an outside lab and like doctors, a second opinion never hurts.

Chapter 21
How should these results be used?

Failure analysis is sometimes difficult because the planned steps cannot be made very far in advance as the results of the first test often determine the best path for the next step. That being said there are many standard paths that are followed. You can, and should, make an "if then" flow for most of the FA issues you see repeatedly. With these standard flows established and known to the lab staff and accepted by your repeat customers, the long down times waiting for a discussion and approval of the next step can be avoided. A quick three-day FA can turn into a couple of weeks if your customer requires their ok before each step. This results in aggravation for everyone.

A normal set of analysis that most will agree to starts with optical inspection. A control unit that appears normal should have images collected along with any abnormalities observed on the other devices. Trust me you will be glad you took these images later when someone asks for proof of any abnormalities or the markings on the device after you have decapped or sectioned it and all the markings are gone. This practice also makes report writing easier and gives a good starting point for anyone reading it.

After this, 2D X-ray image is used to see if anything obvious can be seen. This step is also a proof of what is present before decapping or otherwise disassembling the device. I have done this step and seen issues in devices that are impossible if what my customer had told me was true. Parts that went through full assembly and test and then failed later were found by X-ray to have no die, no wire bonds, dies that are 90 degrees off in placement, wire bonds assembled with the wrong program, die with a large metal strip from corner to corner that shorted out all levels in the die, and die attach material coming up and over the edge of the die shorting all the bond pads.

These issues mean that the part could not have gone through the testing and inspection claimed. So, someone is in trouble and should get fired. So, those people/groups responsible are going to try to say that the FA is wrong and you are sloppy, anything to avoid taking responsibility. To be fair, some groups and individuals do accept responsibility and want to understand the issues to improve the process, many do not.

Often the different parties involved will want to blame each other if not you and this can become a political game. FA should stay out of politics as much as possible. This is why you need to do a step by step process with a thoroughly and fully documented FA. Be prepared. Going to a meeting with images and just saying what you found will not and should not be satisfactory. You are a professional, so act like one and testify with evidence, then step out of the way and let business decisions be made based partially on your results. As they say on Dragnet, "Just the facts Ma'am."

https://doi.org/10.1515/9781501524790-021

This is to protect yourself, expose what the failure actually is, and to help your company in the long run. Companies can spend a lot of time and money on the wrong things if an FA goes sideways. Also, a lot of second guessing can be avoided by sticking to the agreed upon path and documentation with images. People will argue written words and suggest that the wire bond sweep for instance may have happened during the decap process, but an X-ray image showing swept wire bonds before the decap will close such deflections down. It is important for the lab to develop a reputation for honesty, especially when the lab has done something incorrectly; this will help you in the long run. I have always told my staff that, "Confession is good for the soul." I can fix something if I know about it.

The question FA should push is "What happens now that the FA is done and shows a problem?" Often the parties responsible will want to ignore the issue and say it is rare and they will be more careful. This is a sign of poor management and wishful thinking. Since FA is rarely in a position to demand changes, I suggest that suggested corrective actions be provided in the appropriate meetings and reports.

Nothing is more irritating for FA than seeing the same issue show up in the lab every few weeks with no corrective actions taken. I have had an individual ask me repeatedly in meetings if I was "absolutely sure" of a result every time a topic his group did not want to take responsible for came up in FA. Images of the failure could not satisfy him. Eventually, I had to admit that sorcery or alien technologies more advanced than ours could have done what he suggested, but short of that this was his group's defect and they should get started on correction and give up on deflection. Without the images and each step covered in the FA report, I could not have been so firm. Evidence is your friend in FA. You are the neutral provider of facts. The outcome and effect of the results cannot sway you or you become useless.

We have found failure analysis to be an interesting and rewarding field. We are the assistants to technology progress. We figure out what is wrong and can offer insights on how to make things better. I used to tell people that we are the Quincys of the technological world, now that is an old reference and today, I just say we are the Abbies from NCIS. Abbie is of course welcome to visit the lab anytime.

FA results properly used can improve products, make companies more profitable, and help the world become a safer, more interesting, and better place.

References

[1]   https://www.eag.com/resources/whitepapers/surface-and-interface-characterization-of-polymers/

[2]   https://www.eag.com/resources/whitepapers/which-method-of-sectioning-is-best-for-my-sample/

[3]   DellEMC REL0164 A05 Dye and Pry Procedure.

[4]   IPC-TM-650 2.4.53 Test Method Dye and Pull.

https://doi.org/10.1515/9781501524790-022

Appendix

%RH	Percentage relative humidity
AES	Auger electron spectroscopy
ATE	Automated test equipment. Used to run diagnostics on wafers and packaged semiconductor devices
CD	Critical dimension
CT	Computed tomography three-dimensional rendering typically associated with X-ray or C-SAM
C-SAM	C-mode scanning acoustic microscopy also called SAT (typically in Asia)
Decap	The removal of encapsulating material on an IC to expose the die
Dead bug	Device with leads bending up when chip faces up
DB-FIB	Dual beam focused ion beam, also call the cut and look tool. It combined a SEM imaging system with an ion cutting beam
DOE	Design of experiments
DUT	Device under test
EDS/EDX	Energy dispersive spectroscopy/ X-ray
EELS	Electron energy loss spectroscopy
EFA	Electrical failure analysis
ESCA	Electron spectroscopy for chemical analysis (also XPS)
FA	Failure analysis
FC	Flip chip
FIB	Focused ion beam
FTIR	Fourier transform infrared (spectroscopy)
Hot spot	An unfortunately used name for any spot found by EMMI, OBIRCH, IR, and other assorted defect localization tool. It originates from the use of liquid crystal that would turn black when exposed to heat in particular temperature ranges
IC	Integrated circuit
IV curves	Current (I) versus voltage (V) plots
KGU	Known good unit
Leakage	Current that runs between two locations where the current should be zero
OL	Open line, infinite resistance
PCB	Printed circuit board
PFA	Physical failure analysis
P-lap	Parallel lapping or delayering
SEM	Scanning electron microscopy
SOP	Standard operating procedures
SS	Stainless steel
STP	Standard temperature and pressure 1 atm, 25 °C
Street	Area between die on a wafer
TDR	Time domain reflectometry
TEM	Transmission electron microscope
XPS	X-ray photoelectron spectroscopy (also ESCA)
X-section	Cross section

https://doi.org/10.1515/9781501524790-023

List of figures

https://doi.org/10.1515/9781501524790-024

Index

https://doi.org/10.1515/9781501524790-025

www.ingramcontent.com/pod-product-compliance
Lightning Source LLC
Chambersburg PA
CBHW081545220326
41598CB00036B/6569